Lab Manual for use with

# CHEMISTRY   The Molecules of Life

## Trace Jordan
## Neville Kallenbach

With contributions by William Gunderson

New York    Oxford
OXFORD UNIVERSITY PRESS

Oxford University Press is a department of the University of Oxford. It furthers the University's objective of excellence in research, scholarship, and education by publishing worldwide. Oxford is a registered trademark of Oxford University Press in the UK and certain other countries.

Published in the United States of America by Oxford University Press
198 Madison Avenue, New York, NY 10016, United States of America

© 2018 by Oxford University Press

ISBN: 9780199946204

9 8 7 6 5 4 3 2 1
Printed by Webcom, Inc., Canada

# CONTENTS

# PREFACE

We have written this lab manual to accompany *Chemistry: The Molecules of Life*, published by Oxford University Press USA. The focus of the Explorations and their sequence in the manual align with the presentation of topics in the textbook chapters. We believe that a hands-on laboratory experience will enhance student understanding and appreciation of the scientific content in *Chemistry: The Molecules of Life*.

We wish to thank several individuals at Oxford University Press USA for their valuable assistance with preparing this manual. Jason Noe, Senior Editor, kept us on track and provided constant encouragement. Andrew Heaton, Associate Editor, provided guidance on the development of the manual. Micheline Frederick, Senior Production editor was responsible for turning our manuscript into a finished product. In addition, Professor William Gunderson (Hendrix College) enhanced the manual by contributing Explorations 5, 9, 12, and 17. We are grateful to Professors Brian Coppola (University of Michigan) and John Henssler (New York University) for providing the concept for Exploration 1. Finally, we wish to thank Emily McKinstry and John Augello, Senior Laboratory Preparators at New York University, who developed many of the experimental procedures and helped with formatting the manual to ensure a consistent style.

In conclusion, we hope that the Explorations provided in this lab manual will provide students with the important and beneficial experience of *doing* science in addition to reading about it.

Trace Jordan
Neville Kallenbach

---

# EXPLORATION 1
## Who Has the Same Substance I Have?
### An Introduction to Scientific Investigation

---

## Overview

Scientists rely on experiments and observations to investigate the natural world. They also need to consider various factors that affect these observations. In this lab activity, you will perform some experimental tests on an unknown substance. Your goal is to use your experimental observations to figure out who else has the same substance you have. Before performing these procedures, you will have a class discussion to develop a consistent experimental method that allows comparison of your observations.

## Preparation

You should be familiar with the following concepts and techniques:

- Physical properties
- Chemical properties
- Reproducible experiments
- Experimental protocol

## Equipment and Supplies

**Equipment**
Magnifying glass
Waste beakers

**Supplies**
Unknown samples
Test solutions (A, B, C, D, E)
Spot plates (a plate containing 12 wells)
Toothpicks
Plastic pipettes
Label tape
Sharpie pen

# INTRODUCTION

Scientific knowledge is based on a systematic investigation of the natural world. The stages of this investigation are called the **scientific method**. Two important components of the scientific method are performing **experiments** and making **observations**. This lab project will develop your skills at making, recording, and sharing scientific observations.

First, you will observe and record the properties of an unknown solid substance. **Physical properties** are the characteristics you can observe without changing the chemical composition of the substance. Examples of physical properties include color, odor, texture, melting point, boiling point, density, solubility, and so on.

Next, you will investigate what happens when you add a variety of unknown solutions to this solid substance. In some cases, but not all, mixing these substances will generate a chemical reaction. **Chemical properties** are characteristics that arise from a chemical reaction that changes the identity of the substance. Examples of chemical properties include color changes, the ability to burn, the production of a gas, and so on.

Scientific experiments and observations must be **reproducible**. This means that you must be able to reproduce the results of the experiment if you perform it again under the same conditions. If the experiment is not reproducible, your original observation likely was flawed in some way. Reproducibility also means that somebody else must be able to duplicate your experimental results. For this to happen, you accurately record your **experimental protocol** so that it can be shared with other investigators.

In this lab activity, you will first identify the variables that need to be considered as part of an experiment protocol. Next, you will collaborate as a class to develop a consistent experimental protocol that everyone will use. Finally, you will implement this protocol to examine the properties of your sample.

# EXPERIMENTAL PROCEDURES

---

**SAFETY INFORMATION**

**CAUTION:** Wear GLOVES and SAFETY GLASSES during this experiment.

---

In this exploration, **you** will be responsible for designing the experimental procedures.

## Identifying Your Sample

1. You will have *four unknown solids* at your bench. Each student at the bench will chose a *different* sample to investigate. Answer **Question 1** on the Data Sheets to record your sample.

## Physical Properties

1. What are the *physical properties* of your unknown substance? Examine the sample's appearance (color, texture, and so on), and record your observations in **Data Table 1** on the Data Sheets.

## Chemical Properties

1. You will now investigate the chemical properties of your unknown substance. You will add five unknown test solutions (labeled A, B, C, D, and E) and observe any changes that occur. You are provided with the following materials:

   - Unknown sample
   - Test solutions (A, B, C, D, E)
   - Spot plate (a plate containing 12 wells)
   - Toothpicks
   - Pipettes
   - Label tape
   - Sharpie pen
   - Magnifying glass
   - Waste beaker

2. Consult with your lab partner to discuss what **experimental variables** you need to consider when examining the chemical properties of your unknown substance. Answer **Question 2** on the Data Sheets.

3. The goal of this lab investigation is to determine who else has the same substance you have. To compare your experimental results with those of another student group, a standard experimental protocol is needed that takes experimental variables into account.

   Your lab instructor will facilitate a group discussion about the experimental protocol. Use insights from this discussion to design a **standard experimental protocol** that everybody will follow. Answer **Question 3** on the Data Sheets.

4. Apply the standard experiment protocol to investigate the chemical properties of your unknown substance. Record your observations in **Data Table 2** on the Data Sheets.

5. After you have recorded your observations, share your results with everyone else in the lab class. Add your data to the Class Data Table that will be organized by your lab instructor.

## Who Has the Same Substance I Have?

1. According to the Class Data Table, which student has the same substance you have? What is the evidence for your deduction? Answer **Question 4** on the Data Sheets.

2. Collaborate with the student who has the same substance you have, and give a brief presentation of your observations and conclusions to the class.

3. Now that you have completed your experiment and analyzed your results, how would you improve the experimental protocol if you were to run through it again? Answer **Question 5** on the Data Sheets.

# CLEAN-UP CHECKLIST

It is essential that you clean up properly after the experiments are complete.

❑ If you used the **spot plate, waste beaker, or spatula**, wash it at the sink with soap and water. Dry it with a paper towel, and return it to your lab bench.

❑ If you used **toothpicks, pipettes, label tape, paper towels, or gloves**, dispose of these used materials in the trash.

❑ Wipe up any **powder or liquid spills** on your bench with a paper towel, and dispose of this in the trash.

---

# DATA SHEETS:
# Who Has the Same Substance I Have?

---

Name: _____

Lab Section (Day/Time): _____ Lab Instructor: _____

## Identifying Your Sample

**Question 1:** What is your sample? Write the number of your sample here: _____

## Physical Properties

### Data Table 1: Physical Properties

| Observations of Physical Properties | |
|---|---|
| **Sample _____** | |

## Chemical Properties

**Question 2:** What **experimental variables** do you need to consider that may affect the results of your experiment?

**Question 3:** Based on the class discussion, write the **standard experimental protocol** that everyone will use to investigate the chemical properties of the unknown substance. Write this protocol using **numbered steps** (e.g., 1., 2., and so on) with **clear procedures** for each step.

## Data Table 2: Chemical Properties

| Observations of Chemical Properties | | | | | |
|---|---|---|---|---|---|
| | **Solution A** | **Solution B** | **Solution C** | **Solution D** | **Solution E** |
| **Sample \_\_\_\_** | | | | | |

## **Who Has the Same Substance I Have?**

**Question 4:** By looking at the available date, identify which student has the same substance you have. Write the number of the sample below. Provide three examples of observational evidence that you used to make your deduction.

**Sample number of the same substance I have:** _____

Evidence from observations:

1.

2.

3.

**Question 5:** Provide **two suggestions** for improving the experimental protocol used in this activity. Write your answers below.

1.

2.

# EXPLORATION 2
# Extracting Iron from Breakfast Cereal

## Overview

Iron is an essential mineral for many biological functions such as the transport of oxygen. Some breakfast cereals and other food products add iron to help with our intake of this necessary element. In this lab project, you will extract iron from a sample of fortified cereal using a magnet and then confirm the presence of iron using two simple tests.

## Preparation

You should be familiar with the following concepts and techniques:

- Importance of iron for human health
- Recommended Dietary Allowance (RDA) and Percent Daily Value (%DV)
- Conversion of elemental iron into iron ions
- Structure and function of hemoglobin

## Equipment and Supplies

Bring a calculator to this lab session.

| **Equipment** | **Supplies** |
|---|---|
| Hot plate | Cereal, Total |
| Beaker, 1 L | Weigh boat |
| Thermometer | Hydrochloric acid, 1 M |
| Scale | Iron sulfate, 0.01 M |
| Stir plate | Potassium ferricyanide, 0.1 M |
| Stir bar | Test tubes |
| Neodymium magnet | Permanent marker |
| Graduated cylinder, 10 mL | Disposable pipettes |

# INTRODUCTION

## Nutritional Value of Iron

To maintain your health, your body needs a variety of chemical elements. One of these elements is iron, which is found naturally in many foods, such as red meats and leafy, green vegetables. We need iron to survive because it plays an important role in transporting oxygen throughout our bodies. Without enough iron, it is possible to develop **iron deficiency anemia**. This condition leads to many symptoms, such as tiredness, lack of concentration, and reduced immune function.

Many of us do not get enough iron from these natural sources. Consequently, iron is added to some breakfast cereals and other foods to ensure that we obtain our daily dose. The addition of iron is called **fortification**. Figure 1 shows a nutrition label from a fortified cereal. The Recommended Dietary Allowance (RDA) for an average adult is 18 mg of iron each day. Nutrition labels on food products represent how much of the RDA is contained in one serving by using the Percent Daily Value (%DV). For example, a fortified cereal with a 100% DV for iron contains 18 mg in one serving.

| **Nutrition Facts** | **Amount/Serving** | **% Daily Value** |
|---|---|---|
| | **Total Fat** 0.5 g | 1% |
| | Saturated Fat 0g | 0% |
| | Trans Fat 0g | |
| Serving Size 3/4 cup (29g) Servings Per Container 17 | **Cholesterol** 0mg | 0% |
| | Sodium 170mg | 7% |
| **Calories** 90 Calories from Fat 5 | Vitamin A 25% Iron 100% Zinc 100% Thiamin 100% Calcium 0% Copper 6% | |

**Figure 1** Sample nutrition label for a fortified cereal.

## Absorption of Iron

The iron contained in breakfast cereal is called **elemental iron,** which has the chemical symbol Fe (from the Latin word *ferrum*). It is the same form that exists when iron atoms are found in nature as a chemical element. However, elemental iron cannot be used directly in the human body. To be useful, the iron atom must be converted an **ion**. An iron ion, designated as $Fe^{2+}$, is an electrically charged entity that is formed when an iron atom loses two of its electrons.

The conversion of Fe into $Fe^{2+}$ occurs in the acidic environment of the stomach. The liquid contents of our stomach are acidic due to the presence of hydrochloric acid (HCl). As seen in Reaction 1, HCl reacts with Fe to produce $Fe^{2+}$ ions plus $Cl^-$ ions and hydrogen gas. The labels following the chemical formulas indicate whether the substance exists as a solid *(s)*, liquid *(l)*, or gas *(g)* or in an aqueous solution *(aq)*. Once in the ionic form, $Fe^{2+}$ can be absorbed through intestines and used by the body.

$$Fe(s) + 2HCl(l) \rightarrow Fe^{2+}(aq) + 2Cl^-(aq) + H_2(g)$$

**Reaction 1**

## Hemoglobin

Of all the iron in our body, 70% can be found in a protein called **hemoglobin** (Figure 2). This protein is found in red blood cells and is what gives these cells their red color. Hemoglobin is made up of four subunits: two alpha-chains and two beta-chains. Each subunit has a heme group, which is a ring-shaped molecule that contains an $Fe^{2+}$ ion at its center.

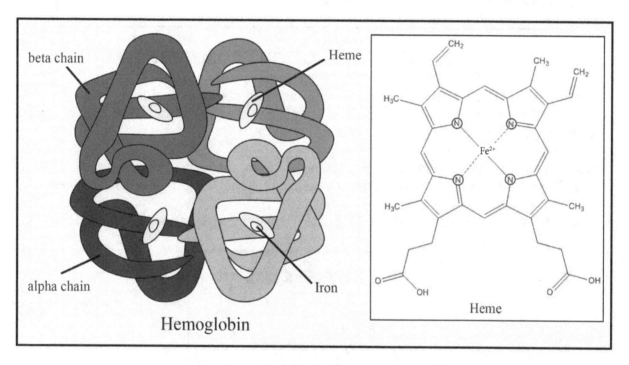

**Figure 2** Hemoglobin and a structural representation of one of its heme groups.

Hemoglobin transports oxygen molecules throughout the human body. When we take a breath, the oxygen molecules that we inhale pass through our lungs into the bloodstream. Hemoglobin can transport a total of four oxygen molecules, because one oxygen molecule attaches to the $Fe^{2+}$ ion at the center of each heme. After carrying the oxygen molecules through the bloodstream, hemoglobin releases them into oxygen-poor environments. The oxygen molecules are then used for chemical reactions within the body's cells, which generate life-sustaining energy.

---

# EXPERIMENTAL PROCEDURES

---

## SAFETY INFORMATION

**CAUTION: The solutions you will be using are dilute, but some are toxic or harmful to your eyes and skin. GOGGLES and GLOVES must be worn for the duration of the experiment.**

## PART A: IRON EXTRACTION FROM CEREAL

1. Take the 1L beaker to the back of the room and add 600 mL of distilled water from the carboy. Place the beaker with water onto a hot plate, and add a thermometer. Turn on the hot plate to its highest setting, and let the water heat up to 50°C.

2. Get a large weigh boat from the empty lab bench. Place the weigh boat on the scale, and hit the "Tare" button. Use the tared weigh boat to measure out 60.0 g (two servings) of cereal.

3. Once the water has reached 50°C, add the cereal and water to the blender. Refer to Figure 3.

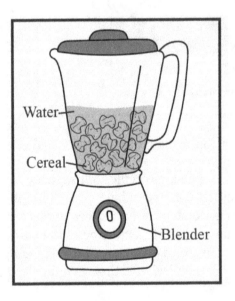

**Figure 3** Creating a cereal/water extract mixture.

4. Make sure that the top of the blender covers the opening near the lip; otherwise, the cereal may come out! Blend the mixture at a high speed for a few minutes until the cereal has been finely ground to a uniform consistency.

12

5. Pour the mixture back into the 1L beaker. If some cereal remains in the blender, add some distilled water from the bottle at your bench to wash the remaining cereal into the beaker.

6. You are now ready to extract the added iron from your cereal sample. Place the 1L beaker onto a stir plate, and add a stir bar. Refer to Figure 4.

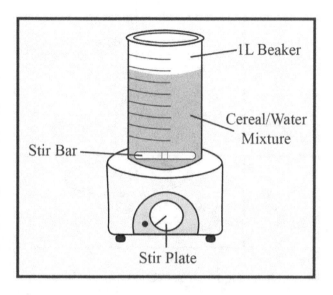

**Figure 4** Iron extraction set-up.

7. Set the stir plate at its highest setting, and start a timer for 10 minutes.

8. Get a small weigh boat from the empty lab bench. While you are waiting, measure the mass of the small weigh boat using a scale that can measure grams to the thousandth place. Record this value in **Data Table 1**. You will use this weigh boat to find the mass of the iron you extract.

9. Once your timer has gone off, turn off the stir plate, and place the beaker on the lab bench. Use the stir bar retriever to remove the stir bar from the beaker.

10. Hold the stir bar retriever and stir bar above the beaker, and use the distilled water bottle to **gently** wash off any cereal. **Make sure you do not wash off the iron filings!**

11. Now you will remove the iron filings from the stir bar using a powerful **neodymium magnet**. Place the neodymium magnet underneath the weigh boat you used in step 8.

    **BE CAREFUL! These neodymium magnet has a very strong attractive force. Keep it at a SAFE distance from other magnetic materials at your bench (e.g., stir bars, stir plates, and other neodymium magnets).**

12. Carefully place the stir bar above the weigh boat (the opposite side of the magnet). Hold the neodymium magnet, and bring the stir bar near it (with the weigh boat in between) until you feel an attractive pull. Iron filings should move from the stir bar onto the weigh boat due to the attractive force of the magnet. Once you remove the iron from one side of the stir bar,

repeat on the other side until all of the iron filings are on the weigh boat. Refer to Figure 5. Repeat again with the stir bar retriever.

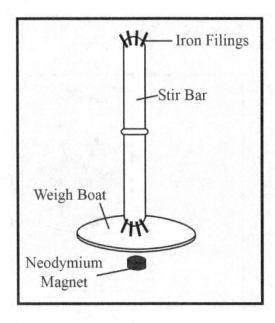

**Figure 5** Collection of iron filings.

13. Place the stir bar and stir bar retriever away from the neodymium magnet on your bench. Use the neodymium magnet under the weigh boat to move the iron filings to a dry area so that you can dab water droplets with a paper towel and dry your sample.

14. After you have removed as much water as possible, place the weigh boat and iron sample on the scale, and record the mass to the thousandth place in **Data Table 1**.

15. Answer **Questions 1 and 2** on the Data Sheets.

## PART B: CHEMICAL TESTS

In this part, you will use the metal filings extracted from Part A in a chemical test to detect the presence of $Fe^{2+}$ ions. The test uses a chemical compound called **potassium ferricyanide**, which has the chemical formula $K_3Fe(CN)_6$.

In this procedure, you will perform two tests, which are called a **negative control** and a **positive control**. The negative control shows the result of the chemical test when no iron is present in the sample. The positive control shows the result of the chemical test when $Fe^{2+}$ ions are added to the sample. These tests provide reference points for comparing our experimental test on the extracted iron.

Table 1 indicates what will be added to each test tube. **Make sure you use a NEW PIPETTE each time to prevent cross-contamination.**

## Table 1: Chemical Tests

| HCl and Filings | Negative Control | Positive Control | Sample |
|---|---|---|---|
| Metal filings (Part A) | 2 mL of $dH_2O$ | 2 mL of $dH_2O$ | 2 mL of $H_2O$ |
| 20 drops of HCl | | 10 drops of $FeSO_4$ | 10 drops of HCl and filings |
| | 1 drop of $K_3Fe(CN)_6$ | 1 drop of $K_3Fe(CN)_6)$ | 1 drop of $K_3Fe(CN)_6)$ |

1. Label a clean test tube as **HCl and Filings**. Use a small spatula to scrape as much of your metal filings into a clean test tube as you can. Try to get all of the filings to the bottom of the test tube and off the sides.

2. Add 20 drops (1 mL) of HCl to the test tube with a disposable pipette. The HCl will mimic the reaction that occurs in the between elemental iron and stomach acid to produce $Fe^{2+}$ ions.

3. Set a timer for 15 minutes to allow the reaction to proceed.

4. While you are waiting, get three more clean test tubes, and label them as **Negative Control**, **Positive Control**, and **Sample**.

5. Measure out 2 mL of distilled water ($dH_2O$) into a graduated cylinder using the bottle at your bench. Add to the distilled water to the **Negative Control** test tube.

6. Add 1 drop of the potassium ferricyanide solution to the **Negative Control** test tube. Record the color of the mixture in **Data Table 2**. Dispose of your pipette into the Waste Beaker at your bench.

7. Measure out 2 mL of distilled water into a graduated cylinder using the bottle at your bench. Add to the distilled water to the **Positive Control** test tube.

8. Collect 0.5 mL of $FeSO_4$ solution using a disposable pipette, and transfer 10 drops of this solution into the **Positive Control** test tube. Dispose of your pipette into the Waste Beaker at your bench.

9. Add 1 drop of the potassium ferricyanide solution to the **Positive Control** test tube. Record the color of the mixture in **Data Table 2**. Dispose of your pipette into the Waste Beaker at your bench.

10. Once your timer has gone off, your reaction in the **HCl and filings** tube will be ready to test for the presence of iron.

11. Measure out 2 mL of distilled water into a graduated cylinder using the bottle at your bench. Add to the water to the **Sample** test tube.

12. Collect 0.5 mL of the liquid in the **HCl and Filings** test tube with a disposable pipette. Transfer 10 drops of this solution to the **"Sample"** test tube. Dispose of your pipette into the Waste Beaker at your bench.

13. Add 1 drop of the potassium ferricyanide solution to the sample test tube. Record your observations in **Data Table 2**. Dispose of your pipette into the Waste Beaker at your bench.

14. Answer **Question 3** on the Data Sheets.

---

# CLEAN-UP CHECKLIST

---

It is essential that you clean up properly after the experiments are complete.

❑ Dispose of the iron-cereal mixture into the sink.

❑ Wash all beakers, blenders, stir bars, stir bar retrievers and spatulas with soap and water, and return them to your lab bench.

❑ Dispose of the **HCl and Filings** test tube into the designated waste container on the back bench.

❑ Pour out the **Sample**, **Positive Control**, and **Negative Control** test tubes in the waste container on the back lab bench. Rinse the test tubes with water, and add the rinse water to the waste container. Rinse enough times until the test tube is CLEAN. Use a disposable pipette to remove any water that remains in the test tube, and dispose of that water in the waste container as well. Place the cleaned and empty test tubes on the rack on the back lab bench.

❑ Throw out all used weigh boats and paper towels into the trash.

❑ Clean up any spills on your lab bench.

---

# DATA SHEETS:
## Extracting Iron from Breakfast Cereal

---

Name: _____ Date: _____

Lab Instructor: _____ Section: _____

## PART A: IRON EXTRACTION FROM CEREAL

### Data Table 1: Mass of Iron

| Mass of Weigh Boat | Mass of Weigh Boat and Iron Filings | Mass of Iron Filings |
|---|---|---|
|  |  |  |

**Question 1:** (a) According to the nutrition label for the cereal box, what is the percent daily value of iron in one serving?

%DV = _____

(b) The Recommended Dietary Allowance for iron is 18 mg. Calculate the total mass of iron in two servings of the cereal.

Mass of iron = _____

**Question 2:** You will likely find that a difference between the mass of iron you extracted (from Data Table 1) and the nutritional data obtained from the food label.

(a) Use the equation below to calculate the **% difference** between your extracted mass of iron and the amount that you calculated in Question 1:

$$\% \text{ difference} = \frac{(\text{label value} - \text{extracted value})}{\text{label value}} \times 100\%$$

% difference =

(b) Propose one reason why the mass of the extracted iron is different from the expected value based on the nutrition label.

# PART B: CHEMICAL TESTS

## Data Table 2: Results of Chemical Tests

|  | Negative Control | Positive Control | HCl and Filings |
|---|---|---|---|
| Potassium ferricyanide test |  |  |  |

**Question 3:** Compare your observations for the chemical test of the negative control, positive control, and the sample of HCl and filings. What do these observations indicate about the presence of $Fe^{2+}$ ions in the test sample?

# EXPLORATION 3
# Investigating Molecular Structure

## Overview

An understanding of chemical bonding and molecular structure in simple molecules is an essential foundation for studying the molecules of life, such as proteins, nucleic acids, lipids, and carbohydrates. This lab project is designed to introduce you to the structure of small molecules and how they can be studied.

## Preparation

You should be familiar with the following concepts and techniques:

- Valence electrons
- Covalent chemical bond
- Lone pairs (also called nonbonding electron pairs)
- Electron pair repulsion and molecular geometry
- Different representations of molecules
- Molecular geometry for methane, ammonia, and water

## Equipment and Supplies

Bring a calculator to this lab session.

**Equipment**
Molecular model kit
Protractor

**Supplies**
Colored pencils
Index cards

# INTRODUCTION

## Molecular Model Kits

During this exploration, you will use a **molecular model kit.** In these kits, the atoms are represented by balls of various colors, and the chemical bonds are represented by gray sticks. Because of these two components, this type of molecular model is called a **ball-and-stick model**. There are two types of sticks: long and short. Use one of the short sticks for single bonds and two of the longer, more flexible sticks to make double bonds, as seen in Figure 1.

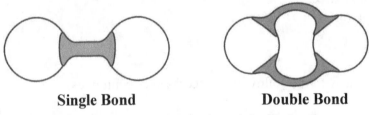

### Single Bond          Double Bond

**Figure 1** How to make single and double bonds.

To assist you with your activities, remember some bonding rules for the common elements H, C, N, and O because they provide useful guidelines for constructing molecules. The number of bonds normally formed by each type of atom is listed in Table 1, as is the color scheme used in the modeling kit.

**Table 1 Colors and Bonding Rules for Common Atoms**

| Atom | Color | Number of Covalent Bonds |
|---|---|---|
| Hydrogen, H | White | 1 |
| Oxygen, O | Red | 2 |
| Nitrogen, N | Blue | 3 |
| Carbon, C | Black | 4 |

## Molecular Structures

Throughout the lab, you will have to draw either the **molecular model** or **two-dimensional (2D) structure** of the molecules you build with the molecular model kits. In Figure 2 shows a visual representation of each type of drawing.

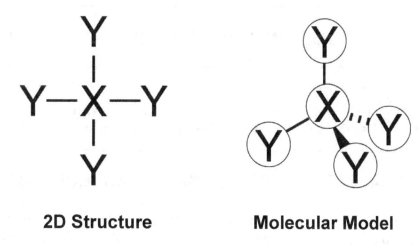

**2D Structure**          **Molecular Model**

**Figure 2**  Examples of drawn structures.

For the **molecular model**, represent the atoms by balls and the bonds by sticks. Illustrate the geometry of the molecule in your drawing. Use the colored pencils to indicate the various atom types, and assign chemical symbols (e.g., C or H) to the atoms. Since hydrogen is white in the model color scheme, indicate these atoms as circles without any shading.

For the **2D structure**, use letters (e.g., C or H) to represent the atoms and straight lines to indicate covalent chemical bonds. A 2D structure is drawn flat on the page, and you do NOT need to capture the 3D geometry of the molecule.

## Molecular Bond Angles

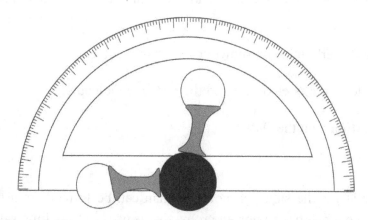

**Figure 3**  How to measure the bond angle of a molecular model.

Lastly, you will have to measure the angle between certain bonds of the molecules you build. Make sure to record the angle that you *measure*, not the angle predicted by theory. The easiest and most accurate way to measure the bond angle is to lay the protractor flat on top of a blank sheet of paper. For example, if we were measure the bond angle of methane ($CH_4$), you would orient one C—H bond along the 0° line and measure the angle for another C—H bond. Refer to Figure 3 above.

# MODELING PROCEDURES

**You will be provide with a POWERPOINT FILE that provides images of various molecular structures. Use these PowerPoint slides as a guide during the exploration.**

## PART A: STRUCTURES OF SMALL MOLECULES

Hydrogen is a very common atom in many molecules, including those found in our cells. Since the hydrogen atom has only a single valence electron, it always forms a single covalent bond with only one atom. The computer gives the structures of three small molecules that are formed when hydrogen atoms form chemical bonds with other atoms:

- Methane, $CH_4$
- Ammonia, $NH_3$
- Water, $H_2O$

### Methane

Methane ($CH_4$, is commonly known as natural gas and is used as a heating fuel in many homes and industries. The methane molecule has a specific three-dimensional geometry that is called a **tetrahedron**; we can equivalently say that the molecular structure of methane is **tetrahedral**.

1. Examine the PowerPoint slide showing a methane molecule.

2. Use your molecular model kit to build the methane molecule.

3. Answer **Question 1** on the Data Sheets.

### Ammonia

Ammonia ($NH_3$) is the simplest molecule than can be formed from nitrogen and hydrogen atoms. It is also a gas and has a pungent aroma that is used in "smelling salts" and in Windex.

1. Examine the PowerPoint slide showing an ammonia molecule.

2. Use your molecular model kit to build the ammonia molecule.

3. Answer **Question 2** on the Data Sheets.

4. **Leave the molecule on your bench for use later in the lab session.**

# Water

Water ($H_2O$) is a familiar molecule and is central to many biological processes. Unlike methane and ammonia, $H_2O$ molecules exist in the form of a liquid instead of a gas at room temperature and atmospheric pressure. This seemingly simple difference is the basis for life on Earth since all living organisms—from humans to the simplest bacteria—require liquid water to survive.

1. Examine the PowerPoint slide showing a water molecule.

2. Use your molecular model kit to build the water molecule.

3. Answer **Question 3** on the Data Sheets.

# Space Filling-Model

The ball-and-stick model of a molecule is useful for showing the types of atoms that exist within a molecule and how they are connected by chemical bonds. However, it does not give an accurate depiction of how much volume in space is taken up by the molecule. This feature is important in designing therapeutic drugs since we need to know whether a drug is too bulky to fit into a receptor or the active site of an enzyme.

1. Examine the PowerPoint slide showing a space-filling model of the water molecule.

We show the volume of a molecule by a **space-filling** representation. This is achieved by considering the **atomic radius** of each atom in the molecule, which is determined by the total number of electrons in the atom. The electrons in an atom take up a certain amount of space around the nucleus, and more electrons will occupy more space. For example, the atomic radius of an element like oxygen, which exists in the first row of the periodic table, will be smaller than the atomic radius of an element like uranium, which has many more electrons that take up space.

The effective size of a molecule can be determined by assigning a **Van der Waals volume** to each atom. This property, which is named after the scientist who determined it, depicts the volume around each atom that causes repulsion with other atoms. A combination of individual Van der Waals volumes for the atoms in a molecule gives the effective space-filling property of the molecule.

2. Answer **Question 4** on the Data Sheets.

## PART B: STRUCTURES OF LINEAR HYDROCARBONS

As their name suggests, **hydrocarbons** are molecules that are composed of only hydrogen and carbon atoms. We have already encountered one hydrocarbon in this lab—methane ($CH_4$). An immense number of hydrocarbons exist, and they adopt a range of structures from long chains to compact rings. We use the term **linear hydrocarbon** to refer to molecules in which the carbon atoms are joined in a "line" like a chain that can vary in length.

Linear hydrocarbons are particularly important as fossil fuels; for example, the primary component of gasoline is **octane**, a hydrocarbon containing eight carbon atoms (hence its name). Hydrocarbons are also found as components within larger biological molecules. This part of the lab will explore various aspects of hydrocarbon structure.

## Ethane

The ethane molecule can be considered as being composed of two —$CH_3$ groups that are joined together, where the "—" line represents a chemical bond to another unspecified atom. The —$CH_3$ grouping of atoms is called a **methyl group** (derived from "methane") and is commonly encountered in many types of molecules.

When you are examining the structure of ethane, note how the hydrogen atoms in each methyl group are arranged. Do the hydrogen atoms on each carbon line up with each other, or do they not?

1. Examine the PowerPoint slide showing an ethane molecule.

2. Use your molecular model kit to build the ethane molecule. Pay particular attention to how the hydrogen atoms are oriented at each end of the molecule.

3. Answer **Question 5** on the Data Sheets.

## Propane

The next hydrocarbon shown in the PowerPoint slides is **propane**, which is commonly used as a portable fuel on camping excursions.

1. Examine the PowerPoint slide showing a propane molecule.

2. Use your molecular model kit to build the propane molecule.

3. Answer **Question 6** on the Data Sheets.

4. **Leave the molecule on your bench for use later in the lab session.**

## Butane

The next molecule shown is **butane**, which is used as the fuel in cigarette lighters.

1. Examine the PowerPoint slide showing a butane molecule.

2. Use your molecular model kit to build the butane molecule.
3. Answer **Question 7** on the Data Sheets.

# PART C: STRUCTURES OF CYCLIC HYDROCARBONS

In addition to forming linear hydrocarbons, carbon and hydrogen atoms can arrange themselves in a variety of ring structures called **cyclic hydrocarbons**.

## Cyclohexane

The first cyclic hydrocarbon is called **cyclohexane** (where "hex" indicates six carbon atoms).

You will observe that each carbon atom makes four single bonds that are arranged in a tetrahedral geometry. Consequently, the six carbon atoms do not lie in the same plane, and the hydrogen atoms protrude both above and below the carbons.

1. Examine the PowerPoint slide showing a cyclohexane molecule.

2. Use your molecular model kit to build the cyclohexane molecule.

3. Answer **Question 8** on the Data Sheets.

## Benzene

**Benzene** is another hydrocarbon with six carbon atoms, but its structure and chemical properties are quite different from those of cyclohexane. Benzene is very stable and is found as a component in a wide range of biological molecules.

The chemical formula of benzene is $C_6H_6$, so it has fewer hydrogen atoms than cyclohexane (and is therefore NOT its isomer). The carbon atoms are found at the vertices of a **hexagon,** and the benzene molecule shows six-fold symmetry.

1. Examine the PowerPoint slide showing a benzene molecule.

2. Use your molecular model kit to build the benzene molecule. To satisfy the bonding rules for carbon, you should build the ring structure with **alternating single bonds and double bonds**.

3. Answer **Question 9** on the Data Sheets.

4. **Leave the molecule on your bench for use later in the lab session.**

Your model of benzene reveals one limitation of the molecular model kits that you are using today. The model can only be built with alternating single and double bonds. In reality, however, studies of benzene have revealed that all of the bonds are equivalent and have a length midway between that of a single bond and that of a double bond. This arises because benzene is more accurately depicted as a **resonance structure**, where the bonding electrons are **delocalized** over the six carbon atoms in the ring.

# PART D: STUDYING A NEW MOLECULE

An important component of this course involves studying new molecules to recognize their chemical structure and configuration. We will practice this skill by investigating a small molecule with a potent biological effect: **amphetamine**. This molecule is stimulant that affects the function of the brain. In addition to being a drug that is often abused, amphetamine is also the basis for drugs to treat attention-deficit/hyperactivity disorder (ADHD).

This exercise shows how larger, more complex molecules can be constructed from combining smaller components. As instructed in earlier exercises, you should have saved the models for three molecules: **ammonia, propane**, and **benzene**. Using the PowerPoint image as a guide, combine these three molecules to build a molecular model of amphetamine. You will need to remove some atoms to make chemical bonds.

1. Examine the PowerPoint slide showing an amphetamine molecule.

2. Use your molecular model kit to build the amphetamine molecule.

3. Answer **Question 10** on the Data Sheets.

Another type of stimulant drug is called **methamphetamine**. It is related to amphetamine and contains an additional methyl group (hence the name "meth"). A **methyl group** (—$CH_3$) is like a methane that is missing one H atom, allowing it to be attached to another atom using the spare covalent bond.

4. Use your molecular model kit to build the methamphetamine molecule. Modify the amphetamine molecule to make methamphetamine by removing one of the H atoms from the —$NH_2$ group and replacing the H atom with a methyl group (—$CH_3$).

5. Answer **Question 11** on the Data Sheets.

# PART E: DEDUCING MOLECULAR STRUCTURES

Knowing the bonding rules for H, C, N, and O—the most common atoms in biological molecules and pharmaceuticals—will enable you to deduce the structures of molecules from their chemical formulas.

In this part, you will be given two index cards that each contain the name and chemical formula of a molecule. As a reminder, a chemical formula tells you the number and type of atoms in a molecule, but it does not tell you how these atoms are bonded together. Your task is to deduce the molecular structure based on the chemical formula and the bonding rules.

1. Examine the index card, and write down the molecule name and chemical formula in **Question 12** on Data Sheets.

2. Build a molecular model that matches the chemical formula, and draw the 2D structure in **Question 12** on your Data Sheets. Remember that the molecule could include single, double, or triple covalent bonds.

3. Repeat the procedure for the molecule on the second index card. Answer **Question 13** on the Data Sheets.

---

# CLEAN-UP CHECKLIST

---

It is essential that you clean up properly after the experiments are complete.

❑ Take apart all of your molecular models. Places the bonds and the individual atoms back in the appropriate compartments.

❑ Scroll back through the PowerPoint presentation to the first slide.

---

# DATA SHEETS:
## Investigating Molecular Structure

---

Name: _____ Date: _____

Lab Instructor: _____ Section: _____

## PART A: STRUCTURES OF SMALL MOLECULES

**Question 1:** (a) Draw the **molecular model** of methane in the box below.

### Molecular Model of Methane

(b) Use the protractor to measure the angle between the C—H bonds in your model.

**Angle:** _____

**Question 2:** (a) Draw the **molecular model** of ammonia in the box below.

### Molecular Model of Ammonia

(b) Use the protractor to measure the angle between the N—H bonds in your model.

**Angle:** _____

**Question 3:** (a) Draw the **molecular model** of water in the box below.

### Molecular Model of Water

(b) Use the protractor to measure the angle between the O—H bonds in your model.

**Angle:** _____

**Question 4:** (a) Draw the **space-filling model** of the water molecule in the box below. Your drawing should capture the relative volumes of the O and H atoms.

**Space-Filling Model of Water**

(b) Does the H atom or the O atom have the larger Van der Waals volume? Explain why.

# PART B: STRUCTURES OF LINEAR HYDROCARBONS

**Question 5:** (a) Draw the **molecular model** of ethane in the box below.

**Molecular Model of Ethane**

(b) Write the **chemical formula** for ethane in the form $C_xH_y$ (where x and y are the appropriate numbers of each atom).

**Chemical formula:** _____

**Question 6:** (a) Draw the **molecular model** of propane in the box below.

**Molecular Model of Propane**

```


```

(b) Write the **chemical formula** for propane in the form $C_xH_y$ (where $x$ and $y$ are the appropriate numbers of each atom).

**Chemical formula:** _____

**Question 7:** (a) For a change, draw the **two-dimensional structure** of butane in the box below.

**2D Structure of Butane**

```


```

(b) Write the **chemical formula** for butane in the form $C_xH_y$ (where $x$ and $y$ are the appropriate numbers of each atom).

**Chemical formula:** _____

## PART C: STRUCTURES OF CYCLIC HYDROCARBONS

**Question 8:** (a) Draw the **two-dimensional structure** of cyclohexane in the box below.

### 2D Structure of Cyclohexane

(b) Write the **chemical formula** for cyclohexane in the form $C_xH_y$ (where $x$ and $y$ are the appropriate numbers of each atom).

**Chemical formula:** _____

**Question 9:** (a) Draw the **two-dimensional structure** of benzene in the space below. In your drawing, indicate which carbon—carbon bonds are double bonds and which are single bonds.

### 2D Structure of Benzene

(b) Write the **chemical formula** for benzene in the form $C_xH_y$ (where $x$ and $y$ are the appropriate numbers of each atom).

**Chemical formula:** _____

# PART D: STUDYING A NEW MOLECULE

**Question 10:** (a) Draw the **two-dimensional structure** of amphetamine in the box below.

### 2D Structure of Amphetamine

(b) Write the **chemical formula** for amphetamine in the form $C_xH_yN_z$ (where $x$, $y$, and $z$ are the appropriate numbers of each atom).

**Chemical formula:** _____

**Question 11:** (a) Draw the **two-dimensional structure** of methamphetamine in the box below. **Circle the methyl group that was added to the molecule.**

### 2D Structure of Methamphetamine

(b) Write the **chemical formula** for methamphetamine in the form $C_xH_yN_z$ (where x, y, z are the appropriate numbers of each atom).

**Chemical formula:** _____

# PART E: DEDUCING MOLECULAR STRUCTURES

**Question 12:** Write down the molecule name and chemical formula from one of the index cards. Draw the two-dimensional structure of the molecule in the box below.

**Molecule Name:** _____     **Chemical Formula:** _____

**Question 13:** Write the molecule name and chemical formula from the second index card. Draw the two-dimensional structure of the molecule below.

**Molecule Name:** _____     **Chemical Formula** _____

# EXPLORATION 4
# Determining the Size of a Molecule

## Overview

Molecules are much too small to be seen with the naked eye, or even with a standard laboratory microscope. So how can we find the approximate size of a molecule?

This dilemma occurs repeatedly in science. We often want to obtain information about entitles such as molecules that we cannot observe directly. In this situation, we must devise an **experiment** in which we can measure **observable properties** and then use them to make deductions about the invisible world.

The experiment you will perform in this exploration is a good illustration of this principle. You will use a drop of oil to create an **oil film** and then measure its area. From this measurement, you will be able to deduce the approximate size of a molecule by making certain assumptions about how the oil film behaves. As good scientists, we will also need think about possible sources of error in the measurements that could affect the accuracy of the final result.

## Preparation

You should be familiar with the following concepts and techniques:

- Why an oil film spreads out over the surface of water
- How the area of an oil film can be connected to the size of a molecule
- Scientific notation and powers of 10
- Scientific units and prefixes such as micro-, nano-, etc.
- How to calculate the size of a molecule (worked example)

## Equipment and Supplies

Bring a calculator to this lab session.

**Equipment**
Finnpipette, 5.0–50 μL
Plexiglass
Aluminum baking dish
Desk lamp

**Supplies**
Chalk
Parafilm
Oleic acid/ethanol solution, 1:200
1% saline solution
Laminated graph paper, 1-cm² grid
Permanent marker

# INTRODUCTION

## What Are Molecules?

Molecules are stable collections of **atoms** that are held together by **chemical bonds**. Their size can range from small molecules with just a few atoms such as water ($H_2O$) to enormously large molecules such as DNA, which contains millions of atoms.

One example of a medium-sized molecule is shown in Figure 1. Called **oleic acid,** it is found as a principal component of vegetable oil. In the figure, the atoms are drawn as "balls" and the chemical bonds that join them as "sticks." The shape of the molecule arises from the specific type of chemical bonding between the atoms.

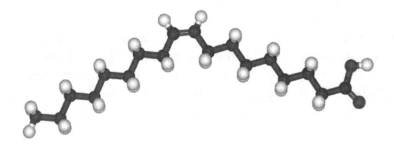

**Figure 1**  A molecule of oleic acid, which is found in vegetable oil.

## How Can We Measure the Size of a Molecule?

When you stroll along a sidewalk after a rain shower, you may notice a thin film of **oil** on the surface of some of the puddles. This oil film reveals itself by how it reflects light differently compared to the surface of the water. The reason you see the oil at all is a manifestation of a chemical rule that you have certainly heard before—**oil and water don't mix**. Instead of being absorbed into the water, the oil molecules float on the surface as a thin film.

We can use this principle as the basis of an experiment to measure the approximate size of a molecule. If you add a drop of oil to the water in a pan, the oil drop will spread out over the surface of the water. A small drop of oil is all that is required to cover a substantial area. It is possible to estimate the size of a molecule if we make one important assumption—namely, **the oil drop spreads out to form an oil layer that is only one molecule thick.**

This assumption is represented in Figure 2. The left side of the figure shows an oil drop with a particular volume $V$, while the right side shows how the oil drop spreads into a thin film with area $A$. Since the same amount of oil is present in each case, we can set up the following equation:

volume of oil drop ($cm^3$) = area of oil film ($cm^2$) × size of 1 molecule (cm)

38

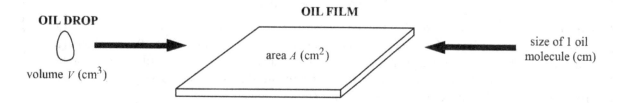

**Figure 2** When placed on the surface of water, an oil drop of volume $V$ spreads out into an oil film of area $A$. We assume that the oil film is only **one molecule thick**.

In the experiment, you will measure the volume ($V$) and area ($A$) to determine the size of a single oil molecule. In practice, you will use a **diluted** solution of oil to obtain an oil film with a reasonable size. You will then examine the assumptions and experimental errors that affect your conclusions about the result. Finally, you will suggest ways to improve the experiment.

To obtain useful results from this experiment, you will need to become familiar with how scientists represent **quantitative information**. These skills will be used in later lab projects and throughout the course.

## Unit Conversions and Calculations

Scientists need to measure and describe dimensions that are very large (like the distance to another star) or very small (like the size of a molecule). This is accomplished by using **scientific notation** to represent numbers and **scientific units** for various types of measurements.

In this experiment, you will be using a small volume unit called a **microliter (μL)**, which is defined a one millionth of a liter. Alternatively, we can say that a million microliters add up to a volume of 1 liter.

$$1 \ \mu L = 10^{-6} \ L \qquad \text{or} \qquad 10^{6} \ \mu L = 1 \ L$$

Another unit used to measure small lengths is called a **nanometer (nm)**. You have probably heard of "nano" in connection with "nanotechnology," which deals with tiny devices on this scale. A nanometer is defined as one billionth of a meter; put another way, 1 billion nanometers fit within a length of 1 meter.

$$1 \ nm = 10^{-9} \ m \qquad \text{or} \qquad 10^{9} \ nm = 1 \ m$$

The box below provides a **worked example** of how you can use your experimental measurements to calculate the approximate size of a single molecule. You will perform a similar calculation using your own data.

## WORKED EXAMPLE

**Question:** You are given a solution that contains oil dissolved in ethanol with a volume ratio of 1:200, respectively. You take a small drop of the solution with a volume of 20 microliters ($\mu$L) and add it to the surface of water in a pan. The oil spreads out to produce a film with an area of 185 cm$^2$. Based on these data, calculate the approximate dimensions of a single oil molecule. Give your answer in units of (a) centimeters and (b) nanometers.

**Answer:** We will divide the calculation into smaller steps.

*STEP 1: Calculate the volume of the oil in units of cm$^3$*

The solution contains an oil:ethanol mix in a volume ratio of 1:200. Therefore,

$$\text{volume of oil } (\mu L) = \text{volume of solution } (\mu L) \times \frac{1}{200}$$

$$= 20 \ \mu L \times \frac{1}{200}$$

$$= 0.10 \ \mu L$$

The area of the film is measured in units of cm$^2$, so we need to convert the volume of oil from $\mu$L to cm$^3$.

**One liter** is defined as the volume of a cube that has a length of 10 centimeters on each side.

$$1 \ L = 10 \ cm \times 10 \ cm \times 10 \ cm = 10^3 \ cm^3$$

We can now convert from $\mu$L to cm$^3$, but we will do it in two steps to make things clearer.

*Convert $\mu$L to L*

$$\text{volume of oil } (L) = 0.10 \ \mu\!\!\!/\!\!L \times \frac{10^{-6} \ L}{1 \ \mu\!\!\!/\!\!L} = 1.0 \times 10^{-7} \ L$$

*Convert L to cm$^3$*

$$\text{volume of oil } (cm^3) = \left( 1.0 \times 10^{-7} \ \not\!L \right) \times \frac{10^3 \ cm^3}{1 \ \not\!L} = 1.0 \times 10^{-4} \ cm^3$$

*STEP 2: Calculate the dimensions of the oil molecule in cm and nm*

If we assume that the oil spreads into a layer that is one molecule thick, we can use the equation from the lab introduction.

$$\text{volume of oil } (cm^3) = \text{area of oil film } (cm^2) \times \text{size of 1 molecule } (cm)$$

We can rearrange the terms in the equation to obtain

$$\text{size of 1 molecule } (cm) = \frac{\text{volume of oil drop } (cm^3)}{\text{area of oil film } (cm^2)}$$

Entering the experimental data gives

$$\text{size of 1 molecule (cm)} = \frac{1.0 \times 10^{-4} \text{ cm}^3}{185 \text{ cm}^2} = 5.4 \times 10^{-7} \text{ cm}$$

Finally, we need to convert this molecular size to units of nanometers (nm). If 1 cm = $10^{-2}$ m and 1 nm = $10^{-9}$ m, then 1 cm = $10^7$ nm. This makes sense because 1 cm is a much larger unit of length compared to 1 nm.

$$\text{size of 1 molecule (nm)} = 5.4 \times 10^{-7} \text{ cm} \times \frac{10^7 \text{ nm}}{1 \text{ cm}} = 5.4 \text{ nm}$$

To give a sense of how small molecules are, we can compare our calculation to the size of a red blood cell, which is the smallest cell in the human body. A red blood cell has a diameter of approximately 8 micrometers (μm), which is equal to 8,000 nm.

---

# EXPERIMENTAL PROCEDURES

---

## Using the Pipette

The success of this experiment relies on your ability to accurately dispense the entire oil volume on the water surface. The following procedures give you practice with using a **pipette**, which is designed to collect and release a precise volume of liquid.

1. You will be using a pipette with an adjustable volume (Figure 3). Here are some important points when using the pipette:

    - You can set the pipette to a specific volume in microliters (μL) by **turning the knob at the top**. Please do this **GENTLY** or else you will damage the pipette.

    - **ALWAYS ADD A PLASTIC TIP** to the end of the pipette for collecting the sample. If you use the pipette without a tip, you will ruin it! Please return the pipette to the stand when you have finished using it; do **NOT** lay it flat on the bench top.

    - The button at the top of the pipette is used for collecting and dispensing the samples. Press the button with your thumb and you will notice that it has **TWO STOPS**. The **FIRST** stop (when your thumb meets some resistance) is used to **COLLECT** the sample. The **SECOND** stop (which requires you to press harder) is used **ONLY** to **DISPENSE** the sample.

**Figure 3** An adjustable pipette used for small volumes (e.g., 20 μL).

2. You have been provided with small strips of **Parafilm** (a waxy plastic) and a colored solution. Use the pipette to transfer separate drops of the solution onto the parafilm with the following volumes. **Turn the knob at the top to obtain different volumes.** You should practice **two times** with **each** volume measurement.

**20 μL**          **15 μL**          **10 μL**

## <u>Making and Measuring the Oil Film</u>

1. Using a **1000 mL beaker**, fill with the salt water solution to the **800 mL line**. **Carefully** take the beaker back to your workbench, and **gently** pour the water into the aluminum dish. You will need to **repeat** this **two times** so that you have about 1.6 L of the salt solution in your dish.

2. Let the water settle for 3 to 5 minutes. You want the surface of the liquid to be as still as possible when adding the oil drop.

3. Set up a desk lamp so that the light from the lamp skims the surface of the water. The light will make it easier to view the spread of the oil.

4. Using the dropper bottle, sprinkle a small amount of chalk powder on the surface of the salt solution. You want to create a **light and even sprinkling** that covers the entire surface of the water and avoids clumps of powder.

5. Add a **new plastic tip** to the pipette, and check that the volume is set to **10 μL.**

6. Use the pipette to collect **10 μL** of the **1:200 oil/ethanol mixture** from the plastic tube.

7. **You will now add the oil drop to the solution**. Bring the pipette tip close to the surface of the liquid, and dispense the oil solution by pushing the top button to the **second stop**. Sometimes the drop will stick to the end of the pipette tip; if that happens, tap the pipette gently with your finger to release the drop.

8. You will notice that the oil immediately spreads over the surface of the liquid to form an **oil film** (the ethanol dissolves in the water). Place the **Plexiglass square** on top of the aluminum dish, and wait for the oil film to stabilize. It should only take a few seconds for the film to stabilize.

9. Once the film has stabilized, place a clear **plastic transparency** on top of the Plexiglass, and use a permanent marker to trace the outline of the oil film. This will involve some judgment about where best to draw the line. **DO NOT WRITE DIRECTLY ON THE PLEXIGLASS!**

10. Place your trace of the oil film on the **graph paper**, which is divided into 1-cm squares (Figure 4). Measure the area of the oil film by counting how many squares are contained within the outline. **Write your result on the Data Sheets.**

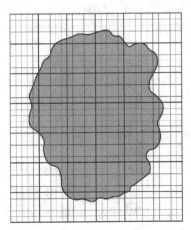

**Figure 4** Using graph paper to measure the area of the oil film.

*HINT*: Use the bigger grid of 25 squares as a guide, and write how many small squares are covered by the oil film within each large grid. You can then add all the numbers to determine the total area of the oil slick.

# CLEAN-UP CHECKLIST

It is essential that you clean up properly after the experiments are complete.

❏ Slowly take the aluminum dish to the sink, and carefully pour all of the contents down the drain. Rinse several times to completely clean all of the chalk powder from the dish. Wipe the dish dry with paper towels. It is important to remove all residual oil and chalk powder from the dish.

❏ Return all supplies to your bench, and double check that the pipette is hanging on the pipette stand.

---

# DATA SHEETS:
# Determining the Size of a Molecule

---

Name: _____ Date: _____

Lab Instructor: _____ Section: _____

## Measuring the Size of an Oil Molecule

Oil: ethanol dilution: _____  Area of oil film ($cm^2$): _____

**Question 1:** Use these experimental data to calculate the size of a single oil molecule. Give your answer in (a) centimeters and (b) meters using scientific notation. Write your answers using **two significant figures** (the appropriate number for the level of certainty for this experiment).

*HINT*: Use the worked example in the lab introduction as a guide.

Remember that

$$1\ \mu L = 10^{-6}\ L \qquad 1\ L = 10^3\ cm^3 \qquad 1\ cm = 10^{-2}\ m$$

**Question 2:** The oil used in this experiment is **<u>oleic acid</u>**, and the length of one molecule is approximately 1.8 nanometers (nm), where $1 \text{ nm} = 1 \times 10^{-9} \text{ m}$.

Convert the molecular size from Question 1 into units of nanometers. How does your calculated size compare to 1.8 nm? Is your calculated size bigger, smaller, or very clos*e* to this quoted value?

**Question 3:** Experiments always involve some degree of **variability**, so it is important to know how much variation can occur. We will now compare the measured sizes from all groups in the lab section.

Write your answer to Question 2 on the Class Data Sheet, and copy **all** the results from the class in the Data Table below. Calculate the **average result** from all these measurements.

### Data Table 1: Comparison of Different Measurements of Molecule Size

| Student Group | Molecule Size (nm) |
|---|---|
| 1 | |
| 2 | |
| 3 | |
| 4 | |
| 5 | |
| 6 | |
| 7 | |
| 8 | |
| 9 | |
| 10 | |
| 11 | |
| **Average** | |

## **Improving the Experiment**

When performing an experiment, we should always be thinking about **sources of error** and about ways to **improve** the experimental procedure. The following questions ask you to analyze the experimental method used in this exploration and think of improvements.

**Question 4:** Identify **two steps** of the experimental procedure that could affect the precision of your calculated molecular size. Would these features tend to produce a calculated value that is too large or too small? Explain you answer for each step.

1.

2.

**Question 5:** Suggest **two ways** that the experimental procedure could be improved to obtain a more precise measurement of molecular size.

1.

2.

# EXPLORATION 5
# Properties and Melting Points of Fatty Acids

## Overview

In this experiment, you will be exploring the physical properties of fatty acids. You will first test to see if a fatty acid is saturated or unsaturated. Then, you will explore the relationship between molecular size and melting point of the saturated fatty acids.

## Preparation

You should be familiar with the following concepts and techniques:

- Molecular structure of fatty acids
- Molecular difference between saturated and unsaturated fatty acids
- Degree of saturation of a fatty acid
- Relationship between the size of a fatty acid and its melting point

## Equipment and Supplies

**Equipment**
Watch glasses
Capillary melting point tubes
Melting point apparatus

**Supplies**
9 Test tubes
Lauric acid
Myristic acid
Palmitic acid
Stearic acid
Linoleic acid
1% bromine solution
methylene chloride

# INTRODUCTION

**Fatty acids** are carbon chains with a carboxyl group at one end and a methyl group at the other. The line structure for two fatty acids, **lauric acid** (found in coconut oil) and **oleic acid** (found in olive oil), are shown in Figure 1. Fatty acids play an important role in many metabolic processes because they can be used as a source of energy for cells. In addition, fatty acids are commonly used in both the pharmaceutical and personal hygiene industries.

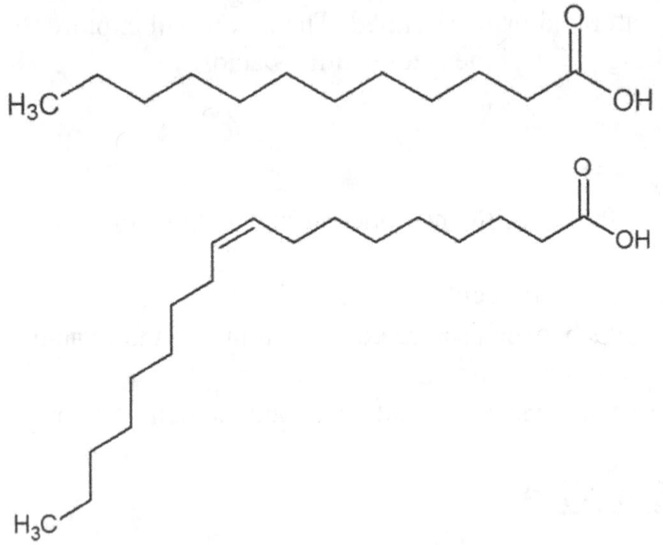

**Figure 1** Line-angle structures of lauric acid (top) and oleic acid (bottom).

Fatty acids are typically 12 to 18 carbon atoms in length and can be either **saturated** or **unsaturated**. A saturated fatty acid has no carbon-carbon double bonds (see lauric acid) and is thus saturated with hydrogen. Unsaturated fatty acids contain one or more carbon-carbon double bonds (see oleic acid). These carbon-carbon double bonds form rigid kinks in the structures of the unsaturated fatty acids. These kinks mean that the acids will not line up well in a crystal and will have a low melting point. At room temperature, unsaturated fatty acids are typically liquids, while saturated fatty acids are usually solid.

In this exploration, you will be given several different fatty acids. You will determine if the acid is saturated or unsaturated. You will also determine the **degree of saturation** (i.e., how many carbon-carbon double bonds are in the unsaturated acids).

After determining which acids are saturated, you will examine the melting point of the different fatty acids. The boiling point of hydrocarbons significantly depends on the molecular size and mass of the compound of interest. The same effect can be seen in the melting points of hydrocarbon compounds. In fatty acids, the same is true as well. Larger fatty acids have higher melting points compared to smaller ones.

To measure the melting point of the saturated fatty acids, you will use a melting point apparatus. This apparatus allows the temperature of a sample to be raised using a heating block. The temperature is measured with a thermometer, and the sample is monitored by observation though a magnifying lens. Several different melting point apparatuses are available. Your lab instructor will direct you in the proper use of the specific apparatus you will be using.

---

# EXPERIMENTAL PROCEDURES

---

## Saturation of Fatty Acids

1. Place a small amount of each provided fatty acid into six separate test tubes. If the sample is a liquid, add five drops of the sample. If the sample is a solid, add small amount (the size of a pinhead) to the test tube. Label each test tube with the fatty acid.

2. To each test tube, add 1 mL of methylene chloride and mix well.

3. Select one of your fatty acid solutions, and add one drop of the 1% bromine solution and shake to mix. If the fatty acid solution turns orange (the color of the bromine solution), stop. If not, continue to add the 1% bromine solution, 1 drop at a time, until the solution turns orange.

4. Record the number of drops required in **Data Table 1** on the Data Sheets.

## Melting Point of Fatty Acids

1. Four of the provided fatty acids (lauric acid, myristic acid, palmitic acid, and stearic acid) are solids at room temperature. Obtain a small amount of each acid, and place it on a watch glass.

2. Using a spatula, carefully crush the acids into a fine powder.

3. For each acid, take a capillary melting point tube and press the open end into the powder, packing some into the tube.

4. Gently tap the bottom of the tube on the bench top to pack the powder into the bottom of the tube. The total sample height should be 1 to 2 mm.

5. Insert the capillaries into the melting point apparatus, and begin heating. Some melting point apparatuses allow up to three samples to be run simultaneously. Heat the sample at a slow rate (~2°C) so that you get an accurate measurement of the melting point.

6. Record both the temperature that the sample begins to melt and the temperature at which it is completely melted in **Data Table 2** on the Data Sheets.

7. Remove the capillary, and discard it in the provided glass disposal container. Turn off the melting point apparatus.

# CLEAN-UP CHECKLIST

It is essential that you clean up properly after the experiments are complete.

❑ All fatty acid solutions should be disposed of in the appropriate waste container.

❑ Rinse all test tubes with deionized water.

❑ Dispose of remaining solid fatty acids in the labeled solid waste container.

❑ Wash watch glasses and spatulas.

---

# DATA SHEETS:
# Properties and Melting Points of Fatty Acids

---

Name: _____ Date: _____

Lab Instructor: _____ Section: _____

## Saturation of Fatty Acids

### Data Table 1: Bromine Addition to Fatty Acids

| Fatty Acid | Drops of 1% Bromine Added |
|---|---|
| Lauric acid | |
| Myristic acid | |
| Palmitic acid | |
| Stearic acid | |
| Oleic acid | |
| Linoleic acid | |

## Melting Points of Fatty Acids

### Data Table 2: Melting Point Measurements

| Fatty Acid | Temperature at Start of Melt (°C) | Temperature at End of Melt (°C) |
|---|---|---|
| Lauric acid | | |
| Myristic acid | | |
| Palmitic acid | | |
| Stearic acid | | |

**Question 1:** The addition of 1% bromine solution indicates the degree of saturation of a fatty acid.

(a) From your results with the 1% bromine solution, which fatty acids are saturated? Which are unsaturated?

(b) Rank the unsaturated fatty acids based on amount of saturation, from **least saturated** to **most saturated**.

**Question 2:** The melting point of a saturated fatty acid depends on the molecular mass. Thus, a fatty acid with a longer carbon chain will have a higher melting point than a fatty acid with a shorter carbon chain.

(a) Which of the saturated fatty acids has the highest molecular weight?

(b) Look up the structures of the four fatty acids for which you measured the melting point. How many C atoms does each have? Based on your data, do you agree that molecular mass affects melting point? Explain your answer.

**Question 3:** The unsaturated fatty acids are all liquids at room temperature. Thus, their melting points are significantly lower than those for the saturated fatty acids. Look up the melting points and structures of the unsaturated fatty acids used in this experiment.

(a) Does the molecular mass have a significant impact on the melting point of an unsaturated fatty acid?

(b) Does the degree of saturation have an impact on the melting point? Explain your answer.

## Improving the Experiment

When performing an experiment, we should always be thinking about **sources of error** and about ways to **improve** the experimental procedure. The following question asks you to analyze the experimental method and think of improvements.

**Question 4:** Identify one step of the experimental procedure that could affect the precision of how you measured your melting points.

# EXPLORATION 6
# Making Esters

## Overview

Since childhood, you have been familiar with the aromas and tastes of fruits, such as pears, oranges, and bananas. These scents are produced by simple organic molecules, many of which are esters.

In this exploration, you will synthesize esters from different alcohols by using acetic acid (the ingredient of vinegar) as the carboxylic acid. Forming esters requires an input of energy to get the reaction started. Heating the reagents usually speeds up a chemical reaction, so we use heat to help start the reaction as well as to accelerate it once the reaction starts. Our synthesis includes a microwave oven as an energy source to further accelerate the reaction. To detect the products, the human nose is a sensitive organ that can distinguish thousands of different aromas.

## Preparation

You should be familiar with the following concepts and techniques:

- Chemical reactions
- Organic acids, alcohols, and esters
- Concentrated as opposed to dilute solutions
- Microwave radiation and its effect on a reaction
- How we sense and discriminate odors in our noses

## Equipment and Supplies

**Equipment**

Microwave oven
  (with a wooden rack to hold 10-cm tubes)
Hot water bath
  (with a rack to hold 10-cm tubes)

**Supplies**

Test tubes, 10 mL in volume
Test tubes, 10 cm
Glacial acetic acid, 100 mL, in a glass bottle
Ethyl alcohol
*n*-Propyl alcohol
*iso*-Amyl alcohol

59

# INTRODUCTION

## The Chemistry of Carbon

What makes carbon so exceptional in the periodic table is its ability to form stable chemical bonds with itself and with many other atoms, leading to an almost innumerable number of different molecules that we call **organic molecules**. Tens of millions of small organic molecules are known and have been classified, and thousands more are added each day to compendia such as PubChem.

Organic molecules, like all molecules, are products that result from chemical reactions. A large number of chemical reactions have been studied in efforts to synthesize complex natural products and to design new drugs for medicinal applications. Our exploration can only touch on one tiny aspect of this huge subject. However, the exploration shows how a large number of different products can be obtained by using one chemical reaction with a series of different reactants. Thus, one reaction effectively can produce a collection of products, making it possible to generate diverse compounds with different properties.

## Formation of Esters

The formation of an ester is show in Figure 1. This equation includes two R groups (R and R•). They denote any of a large variety of groups, such as methyl (—$CH_3$), ethyl (—$C_2H_5$), octyl (—$C_8C_{17}$), and so on. The two R groups can be the same or different, so a huge number of different esters are possible.

**Figure 1** Synthesis of an ester molecule.

How do esters produce their characteristic aromas? First, by losing the polar acid and alcohol groups present in the reactants, ester products tend to be less soluble and more weakly linked to polar water molecules than the starting materials. With weaker forces connecting them to the solvent (i.e., water), esters tend to evaporate and hence are more volatile than either organic acids or alcohols. The characteristic aromas of esters are recognized by receptors in our noses and mouths. While we associate particular esters with fruits, such as apples or bananas, it is important to realize that the ester contents of different fruits are not simply single molecules produced by a given fruit. Instead different varieties of fruits contain blends of different esters, and these can differ from one strain to another.

Some of this information is shown in an ingenious chart devised by James Kennedy, an Australian chemist (Figure 2). Kennedy's chart lists the formulas of more than a hundred esters along with the aromas of the fruits and products that are associated with each. So apples, for example, include at least nine different esters, ranging from propyl methanoate to ethyl decanoate. Different varieties are sure to differ in their relative content of esters, so the ester content of fruits, as well as liquids such as vinegar, wine or even whisky, is complex.

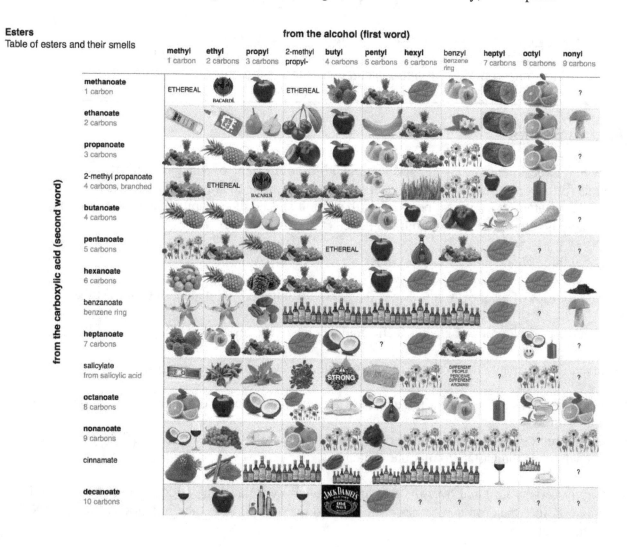

**Figure 2** James Kennedy's chart showing the composition of esters present in a variety of fruits, flowers, leaves, and other sources. **Ethereal** denotes smelling like ether. While only around 150 combinations are shown, our noses can detect many more aromas based on esters, which are only one class of compounds accessible to our array of receptors.

---

# EXPERIMENTAL PROCEDURES

---

### SAFETY INFORMATION

**CAUTION: Please be sure to bring lab safety goggles and to cover your hands with gloves while carrying out these experiments. The acid solution used in these reactions can cause severe burns.**

## Making Esters

In this experiment, you will mix acetic acid, a component of vinegar, with three different alcohols to form a series of different esters that have familiar fragrances. The acetic acid used is very concentrated, so you must be use it with caution and wash any spilled acid with cool water. Acetic acid is referred to as **ethanoate** in the chart shown in Figure 1.

1.  Label three clean, dry, 10-cm test tubes with the letters A, B and C.

2.  Add 6 drops of alcohol A in test tube A, alcohol B in tube B, and alcohol C in the tube C.

3.  Carefully add 2 drops of glacial acetic acid to each of the test tubes.

4.  Insert the tubes into the hot water bath for 3 minutes to initiate the reaction.

5.  Place the tubes into the rack in the microwave oven, and run the oven for 2 minutes on high.

6.  Gently remove the tubes from the oven, and stand them in cool water. Add 20 drops of water to each tube, and agitate to mix the contents.

7.  Gently wave your hand over the tube until you can sense a distinct odor. It may take several waves before the aroma becomes clear. Record the odor of each tube in **Data Table 1** on the Data Sheets.

## Detecting the Aromas of Esters

In the chart shown in Figure 2, nine different esters are associated with apples. However, our nose is capable of distinguishing the smell of apples as apples and not pears. To test our ability to detect mixtures of esters, we now mix equal-sized drops containing mixtures of the ester products together and try to detect how these differ in aroma from the original esters. Report your results in **Data Table 2** on the Data Sheets.

# CLEAN-UP CHECKLIST

It is essential that you clean up properly after the experiments are complete.

❑ Rinse the contents of all test tubes, and return them to the rack.

---

# DATA SHEETS:
# Making Esters

---

Name: _____ Date: _____

Lab Instructor: _____ Section: _____

## Making Esters

**Question 1:** Complete Data Table 1 using the data from your experiments.

### Data Table 1: Making Various Esters

| Carboxylic Acid | Alcohol | Ester product | Fragrance |
|-----------------|-----------|---------------|-----------|
| Acetic acid | *n*-Propyl | | |
| Acetic acid | *iso*-Amyl | | |
| Acetic acid | Octyl | | |

**Question 2:** Naming esters follows a simple rule. For example, the first product would be called *n*-propyl acetate. Use this rule to identify the other two products. How do your names match those in Figure 2?

**Question 3:** If the carboxylic acid were butyric acid instead of acetic acid, what would each of the esters be called?

**Question 4:** Write chemical equations for each of the reactions you have carried out, drawing the structure of the reactants and products as clearly as possible.

(a) Reaction A

(b) Reaction B

(c) Reaction C

## Detecting the Aromas of Esters

**Question 5:** In Data Table 2, identify the type of fragrance that you detect from each of the ester mixtures.

### Data Table 2: Detecting the Fragrance of Ester Mixtures

| Mixture | Fragrance |
|---|---|
| I + II | |
| I+III | |
| II+III | |
| I+II+III | |

**Question 6:** How does mixing the esters affect their fragrance? Make some general observations based on those you recorded in Data Table 2.

## Esters and Soaps

In addition to simple esters such as the ones you synthesized in this exploration, soaps and fats are also esters. Instead of acetic acid, these products rely on long chain acids called **fatty acids**. One example, myristic acid, is shown below:

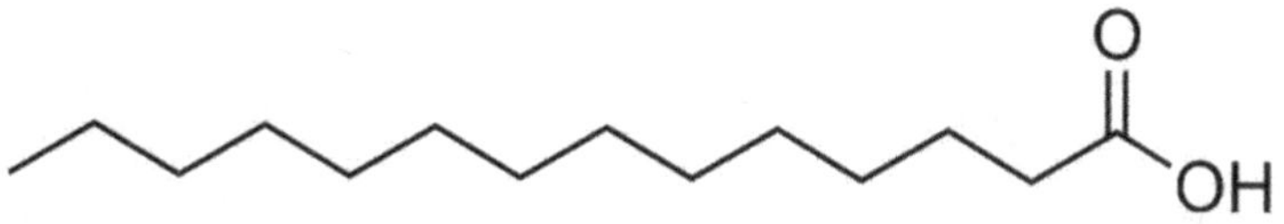

**Question 7:** Do you predict that esters formed by fatty acids will have easily recognizable aromas? Can you think of an explanation in view of our discussion of volatility?

# EXPLORATION 7
# Fermentation in Yeast

## Overview

In this exploration, you will investigate a chemical reaction performed by a living organism. This reaction is called **fermentation**, and it is used by yeast to convert glucose sugar into alcohol. This property of yeast has been harnessed by humans for thousands of years to make beer and wine. Here, we study fermentation using different sugars as a fuel source, measuring production of carbon dioxide ($CO_2$) gas by the cells in a device called a **fermentation tube**.

## Preparation

You should be familiar with the following concepts and techniques:

- Difference between aerobic and anaerobic reactions
- Chemical steps in a fermentation reaction
- Adenosine triphosphate (ATP) as a chemical energy source
- Apparatus used to detect carbon dioxide

## Equipment and Supplies

Bring a calculator to this lab session.

**Equipment**
Erlenmeyer flasks, 50 mL, 100 mL
Graduated cylinders, 100 mL, 25 mL
Fermentation tubes, 100 mL (5 per group)

**Supplies**
Cotton stoppers
Glucose
Table sugar (sucrose)
Lactose
Fructose
Mannose
Sodium fluoride
Baker's yeast
Wax pencil

# INTRODUCTION

Life exists in an almost unbelievable variety of conditions, ranging from deep under Antarctic ice to volcanic hot springs to deep ocean vents as well as gas-containing rock formations miles underground. All organisms require a source of energy to sustain their ability to grow, divide, and develop. Organisms that use oxygen to carry out cellular respiration are the most efficient at generating energy. Mammals, insects, fish, and many kinds of micro-organisms, including bacteria and fungi, utilize oxygen. A chemical reaction that uses oxygen is called **aerobic**, and organisms that use oxygen are called **aerobes**.

Many organisms, however, live in environments that do not provide access to oxygen. Under these conditions, organisms must use **anaerobic** reactions that do not utilize oxygen. One type of anaerobic reaction is called **fermentation**. By definition, fermentation is the conversion of various types of fuel molecules, including sugars, into other molecules, along with the release of chemical energy, in the absence of oxygen.

Suppose the environment of an organism provides it access to oxygen at some times but not others, depending on the circumstances. In this type of environment, it makes sense for an organism to have the ability to accommodate both an aerobic and an anaerobic lifestyle. If oxygen is available and efficient growth is possible, then the organism will use oxygen. But if oxygen is depleted, the organism will use fermentation. Organism that are able to do this include **yeasts**, which are tiny, single-celled fungi that have been used for thousands of years to produce bread, beer, and wine for human consumption (Figure 1).

**Figure 1** A highly magnified microscope image of baker's yeast.
*Source: Mogana Das Murtey and Patchamuthu Ramasamy*

How do simple yeast cells carry out fermentation? First, they use a set of chemical reactions called **glycolysis** (literally "splitting glucose") to convert the glucose sugar into a small organic acid called **pyruvic acid** (Figure 2). Glycolysis involves a series of 10 reactions that extract chemical energy from glucose. Instead of releasing the energy in the form of heat, which would be wasteful, the energy is harvested chemically in the form of **adenosine triphosphate (ATP)**

70

molecules. During glycolysis, glucose, which contains six carbon atoms, is split to form two pyruvic acid molecules, which each contain three carbon atoms). This chemical change generates two molecules of ATP, which are essential for cell growth and division. In fact, ATP serves as a universal carrier of energy that drives most of life's processes. The second stage of the reaction converts the two pyruvic acid molecules into two ethanol molecules, but without the generation of any further ATP. Overall, fermentation generates only two ATP molecules. This product of energy is much less efficient compared to aerobic reactions, but it is sufficient to sustain the survival of the yeast.

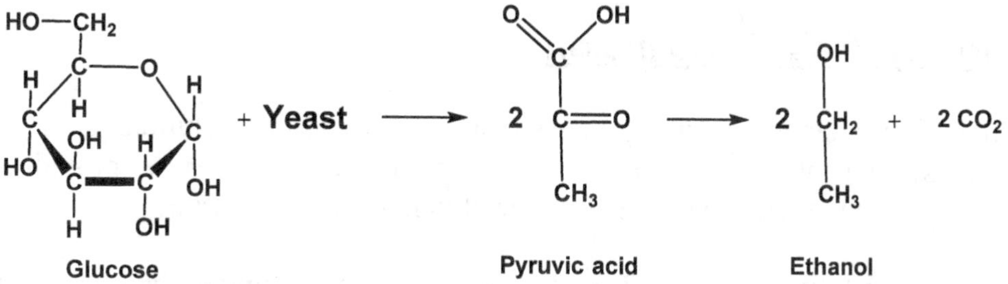

**Figure 2** The chemical steps of fermentation. During the first state of the reaction, glycolysis, the yeast decomposes glucose sugar into two molecules of pyruvic acid. During the second stage of the reaction, these pyruvic acid molecules are converted into ethanol.

In this exploration, you will use yeast cells (supplied as a dry powder utilized for baking) along with several sugars to demonstrate the process of fermentation. You will monitor the fermentation reaction by collecting the carbon dioxide gas ($CO_2$) that is produced by the yeast cells. Questions we can ask about this important process are: Can any sugar provide adequate nutrient for fermentation? How much sugar is needed? And once a cell produces alcohol, is the alcohol tolerated by yeast?

The most direct way to monitor fermentation is to collect the gas released by the yeast cells. To perform this measurement, we use a specialized container called a **fermentation tube** (Figure 3).

**Figure 3** A fermentation tube for measuring release of $CO_2$ by fermenting cells. Closing the neck at the right prevents air from entering the tube, which provides an anaerobic environment for fermentation.

# EXPERIMENTAL PROCEDURES

In the introduction, we explained how yeast use glucose as a fuel for fermentation reactions. You will monitor this reaction by measuring the amount of $CO_2$ produced by the fermenting yeast. Can yeast also perform fermentation using other sugars as a fuel—for example, sucrose, lactose, fructose, or mannose? You will investigate this question in the following exploration.

## Preparation of Yeast Suspension

You will study fermentation using **glucose** plus **one other** sugar solution. Your lab instructor will assign this second solution. In the following experimental protocol, we provide instructions for making every solution. However, you only need to make **two solutions**.

1.  Using a wax pencil, label two 50 mL Erlenmeyer flasks according to the type of sugar added:

    |   |   |
    |---|---|
    | **G** | Glucose |
    | **S** | Sucrose |
    | **L** | Lactose |
    | **F** | Fructose |
    | **M** | Mannose |
    | **N** | Glucose plus sodium fluoride (NaF) |

2.  Add 1 g of dry baker's yeast to each Erlenmeyer flask.

3.  Add 10 mL of warm water to each flask containing the yeast.

4.  Mix each flask thoroughly; allow the flask to stand at room temperature for 5 minutes.

## Preparation of Sugar Solutions

1.  Label two 100 mL Erlenmeyer flasks according to the type of sugar added:

    |   |   |
    |---|---|
    | **G** | Glucose |
    | **S** | Sucrose |
    | **L** | Lactose |
    | **F** | Fructose |
    | **M** | Mannose |

2.  Prepare the following sugar solutions in the associated labeled beaker:

    Dissolve 2 g of glucose in 100 mL of water in flask **G**
    Dissolve 2 g of table sugar in 100 mL of distilled water in flask **S**
    Dissolve 2 g of lactose in 100 mL of water in flask **L**
    Dissolve 2 g of fructose in 100 mL of water in flask **F**
    Dissolve 2 g of mannose in 100 mL of water in flask **M**

## Preparation of Fermentation Tubes

1. Label two 100 mL fermentation tubes according to the type of sugar solution:

   **G**    Glucose
   **S**    Sucrose
   **L**    Lactose
   **F**    Fructose
   **M**   Mannose

2. Add 90 mL of each sugar solution to its associated fermentation tube.

3. Add 10 mL of each yeast suspension to its associated fermentation tube.

4. Stopper the tubes with cotton to prevent air from reaching the contents of the tubes.

## Recording Data

1. After mixing the contents of the tubes, record an initial reading for each tube in **Data Table 1** on the Data Sheets.

2. Monitor the tubes at intervals of **5 minutes** to observe the gas that has evolved. For both tubes, record the volume of gas in milliliters (mL) in **Data Table 1** on the Data Sheets.

3. After 30 minutes, collect the last readings to obtain the full data series. Record these readings in **Data Table 1** on the Data Sheets.

4. Using the **grid** provided in the Data Sheets, plot a graph of the volume of $CO_2$ at each time interval for your two experiments. Add appropriate axes to the graph, and label it with a title. Use colored pencils to distinguish the different data sets.

5. Once the data is graphed, answer **Questions 1 and 2** on the Data Sheets.

---

# CLEAN-UP CHECKLIST

---

It is essential that you clean up properly after the experiments are complete.

☐ Rinse the contents of all the tubes and glassware into the container by the sink using distilled water. Do not pour the solution down the drain.

☐ Remove any labeling from the glassware.

☐ Return all supplies to your bench, and double check that the all supplies are accounted for.

# DATA SHEETS:
## Fermentation in Yeast

Name: _____ Date: _____

Lab Instructor: _____ Section: _____

### Data Table 1: Production of $CO_2$ Over Time

| Fermentation Tube | Initial Reading | Time Interval (Min) | | | | | |
|---|---|---|---|---|---|---|---|
| | | 5 | 10 | 15 | 20 | 25 | 30 |
| Glucose | | | | | | | |
| Sucrose | | | | | | | |
| Fructose | | | | | | | |
| Lactose | | | | | | | |
| Mannose | | | | | | | |

**Graph Title:** _____

**Question 1:** (a) Based on your graph, compare the amount of $CO_2$ produced by the yeast using each of the two sugar solutions.

(b) What do these results tell you about the ability of the yeast to use the second sugar you tested (not glucose) as a fuel source for fermentation?

**Question 2:** Yeast can ferment sugars to produce alcohols, but beyond a certain level, the alcohol becomes toxic to cells. Propose a hypothesis for why alcohol is toxic to cells.

# EXPLORATION 8
# Measuring the Calorie Content of Foods

## Overview

Our bodies and our brain need a constant supply of **energy** to survive. We derive this energy from the food in our diet. In this lab project, you will determine the energy content of two types of food by burning them and measuring the amount of heat that is released. This heat energy is measured in a unit called a **calorie**, which is also used to quantify the energy content of foods. After this exercise, you will use Internet resources to estimate the *total calorie content* of your diet for one day.

## Preparation

You should be familiar with the following concepts and techniques:

- Definition of a calorie as a unit of heat energy
- Three main types of food in our diet
- Adenosine triphosphate (ATP), the energy currency of cells
- Dietary calorie
- Using a calorimeter to measure the energy content of foods
- Heat energy lost to the surroundings

## Equipment and Supplies

Bring a calculator to this lab session.

**Equipment**
Scale, readability of 0.001
Ring stand and clamps
Graduated cylinder, 100 mL
Thermometer
Tongs

**Supplies**
Food samples
Strainer
Soda can
Aluminum foil
Grill lighter
Permanent marker

# INTRODUCTION

## Energy Content of Foods

The food we eat supplies the energy that powers our cells, organs, and bodies. Ideally, we would eat just the right amount of food to provide the necessary energy we require. However, it is sometimes difficult to match our diet to our lifestyle, so we end up eating too little or too much. This lab project will help you gain a better appreciation of the relationship between food and energy.

The energy content of food is measured in **calories**, which is a unit of heat energy. The calorie content of a food is determined by burning it in air and measuring how much heat energy is generated. The calorie is defined in terms of the properties of water:

**One calorie is the amount of heat energy required to raise the temperature of 1 gram of water by 1°C.**

When we go to the gym, we often talk about "burning" carbs or fats. Of course, we don't burn our food in the same way that we burn something in a flame. Instead, our cells use a complex series of chemical reactions to convert the energy stored in chemical bonds into the energy currency of cells—a molecule called **adenosine triphosphate (ATP)**. The ATP molecules are then used to provide the energy for a wide range of cellular process.

The food in our diet can be divided into three main types:

- **Proteins**
- **Carbohydrates**   (also called sugars)
- **Fats**   (also called oils or lipids)

Each of these foods has a characteristic energy content, which derives from the particular types of chemical bonds the food contains. Because the calorie content of food is high, nutrition labels use a unit called a **dietary calorie**—written as a **Calorie** (with a big C).

**1 Calorie = 1000 calories = 1 kilocalorie**

Another way to think of this unit is that 1 Calorie is the heat energy required to raise the temperature of 1 kilogram (2.2 pounds) of water by 1°C (since 1 kilogram = 1000 grams). That's a lot of energy!

The energy content of foods is usually listed in terms of the Calories contained within 1 gram of food:

- **Proteins**        4 Calories/gram
- **Carbohydrates**   4 Calories/gram
- **Fats**            9 Calories/gram

Note that the calorie content of fats is **more than twice** as large as the calorie content of proteins or carbohydrates.

## Measuring Food Calories

We can measure the calorie content of foods using a device called a **calorimeter** ("calorie" relates to heat, and a "meter" is a measurement device). The technique of using a calorimeter to measure heat energy is called **calorimetry**. In this method, the heat energy generated by burning the food is transferred to water contained in an insulating container. The heat energy is absorbed by the water, which **raises its temperature**. From the definition given above, we know that it takes **1 calorie** of heat energy to raise the temperature of 1 gram of water by 1°C. By measuring the mass of the water and its increase in temperature, we know how many calories the water has absorbed.

The experimental equipment you will use is illustrated in Figure 1. A simple soda can will be our calorimeter because it is an enclosed container that permits the transfer of heat energy to the water inside. The temperature of the water will be measured using a thermometer.

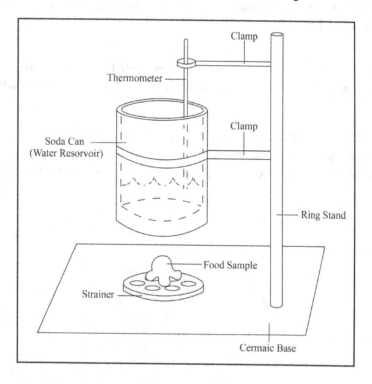

**Figure 1** A diagram of a calorimeter for measuring the heat released when burning food. The food will be placed on the strainer below the calorimeter.

The measured change in temperature of the water is written as $\Delta T$ (where the Greek letter "Delta" indicates "change"). We can then calculate the amount of heat energy absorbed by the water to produce this temperature change.

$$\textbf{energy absorbed by water (cal)} = \textbf{mass (g)} \times \Delta T \textbf{ (°C)} \times \textbf{1 cal/g °C}$$

The "1" at the end of the equation represents the specific heat capacity of water, and its presence is necessary to balance the units. Numerically, however, the total calories absorbed by the water will be determined by the mass of the water and the temperature increase.

In practice, it is often more convenient to measure the **volume** of water rather than **mass**. Fortunately, there is a simple relationship between the volume (in milliliters) and the mass (in grams):

$$\textbf{1.00 mL of water has a mass of 1.00 g}$$

The heat energy absorbed by the water is generated by the combustion of the food. But not all of the combustion energy is absorbed—some of it is lost to the surroundings. For example, a certain fraction of the heat energy is absorbed by the soda can and not by the water inside it (we know this is happening because the can becomes hot to the touch).

Because **energy is conserved**, we can write the following relationship in which all quantities of heat energy are measured in calories:

$$\left[ \begin{array}{c} \textbf{heat of combustion} \\ \textbf{of fuel (cal)} \end{array} \right] = \left[ \begin{array}{c} \textbf{heat energy absorbed} \\ \textbf{by water (cal)} \end{array} \right] + \left[ \begin{array}{c} \textbf{heat energy lost to} \\ \textbf{surroundings (cal)} \end{array} \right]$$

Although the principle of calorimetry is relatively simple, various types of **heat loss** make it difficult to obtain accurate measurements using this method. After performing your measurements, you will be asked to provide suggestions for improving the accuracy of the calorimetry experiment.

# EXPERIMENTAL PROCEDURES

## SAFETY INFORMATION

<u>CAUTION</u>: There will be open flames in this lab. Your lab instructor will point out the location of the fire extinguisher. All students must tie back long hair and tuck away loose clothing. Goggles must be worn for the duration of the experiment.

## PART A: MEASURE THE CALORIE CONTENT OF POPCORN

### Measure the Initial Mass of Food

1. Put one piece of popcorn on the strainer. Measure the mass of the popcorn and strainer using the scale. **Close the cover** of the scale to get an accurate reading.

2. Write the mass in **Data Table 1**, under Trial 1, on the Data Sheets. This value is the initial mass of the food sample.

### Set Up the Calorimeter

1. Place the popcorn and strainer on the white ceramic base of the ring stand. Adjust the position of the popcorn so that it is **underneath** the base of the soda can.

2. Adjust the soda can in the clamp so that it is approximately **1 inch** above the popcorn.

3. Using a 100 mL graduated cylinder, measure out **100 mL of distilled water,** and use a small funnel to pour the water into the soda can.

4. Lower the clamped **thermometer** into the water, making sure that the sensor does not touch the bottom or the sides of the soda can.

5. To minimize heat loss, you want to cover the entire system with aluminum foil (but allow an opening at the bottom to light the food sample and watch the flame). Cover any small openings at the top of the calorimeter with aluminum foil to obtain the best results.

### Recording Temperature Data

1. Before lighting your food sample, note the initial temperature of the water ($T_i$), and record it in **Data Table 1**.

2. **Light the popcorn** by placing the grill lighter underneath it. It may take a little while for the popcorn to catch fire, so continue lighting until you see a flame. As soon you see a sustained flame, put out the lighter.

3. Once the flame has extinguished, note the final temperature of the water ($T_f$), and record it in **Data Table 1**. Calculate the **difference in temperature** ($\Delta T$), and enter it in the table as well.

## Measure the Mass of the Food After Burning

1. When the burnt food has cooled, use the forceps to transfer the food sample and the strainer to the scale. Remember to **close the cover** of the scale to get more accurate results. Record the final mass ($m_f$) of the burnt food that remains, and record the value in **Data Table 1**.

2. Calculate the **difference in mass** ($\Delta m$) between the initial and final masses. Record your data in **Data Table 1**.

## Reset the Experiment

1. You will have to reset the calorimeter before you begin the next experiment. **Be careful touching the soda can because it may still be hot. WEAR HOT GLOVES WHENEVER TOUCHING THE SODA CAN.**

2. Loosen the clamp that holds the thermometer, and move it out of the way.

3. Remove the soda can and the top part of the aluminum cover. Empty the water into the sink.

## Repeat the Experiment

1. Repeat the steps above to measure the calorie content of another piece of popcorn. Add your data from this trial to **Data Table 1**, under **Trial 2**.

2. If the results of your two trials are close, average your results, and record this value in **Data Table 1**. If you have one result that is an "outlier"—that is, it is very different from the other (by, say, a factor of 2)—do not include it in your average. Instead, perform a third trial, and average the two closest results.

## Calculations

1. You now have all the data that you need to calculate the **heat energy absorbed by the water** in the calorimeter. This value provides an estimate of the heat energy released by burning the popcorn. Follow the step-by-step procedures on the Data Sheets.

2. Answer **Questions 1 through 5** on the Data Sheets.

## PART B: MEASURE THE CALORIE CONTENT OF ANOTHER FOOD ITEM

We have provided you with a selection of food items – Cheetos, Doritos, or Fritos. Each student pair will now pick a **second food item** to study via calorimetry. You can choose whatever item you wish to study.

### Repeat the Experiment with the New Food Item

1.  Repeat the experimental procedures from Part A to measure the heat energy absorbed by water when you burn this new food item. For some of the food items, it works best to tilt them at an angle before lighting them with the gill lighter.

2.  Follow the procedures for estimating the calorie content of your chosen food, and enter your data from this experiment in **Data Table 2** on the Data Sheets.

3.  Run through two trials of your second food item to obtain more accurate results. If the results of your two trials are close, average your results, and record this value on **Data Table 2**. If you have one result that is an "outlier"—again by, say, a factor of 2—do not include it in your average. Instead, perform a third trial, and average the two closest results.

4.  Answer **Questions 6 through 10** on the Data Sheets.

---

# CLEAN-UP CHECKLIST

---

It is essential that you clean up properly after the experiments are complete.

❑ Empty the water from the soda can into the sink, and replace the can on the setup.

❑ Wipe down the strainer and ring stand with a paper towel.

❑ Discard all burnt food samples and used paper towels into the trash.

## DATA SHEETS:
## Calorie Content of Foods

Name: _____ Date: _____

Lab Instructor: _____ Section: _____

## PART A: MEASURE THE CALORIE CONTENT OF POPCORN

### Data Table 1: Mass and Temperature Changes for Popcorn

| Trial 1 | | |
|---|---|---|
| **Initial Mass ($m_i$)** | **Final Mass ($m_f$)** | **Change in Mass ($\Delta m$)** |
| | | |
| **Initial Temperature ($T_i$)** | **Final Temperature ($T_f$)** | **Change in Temperature ($\Delta T$)** |
| | | |

| Trial 2 | | |
|---|---|---|
| **Initial Mass ($m_i$)** | **Final Mass ($m_f$)** | **Change in Mass ($\Delta m$)** |
| | | |
| **Initial Temperature ($T_i$)** | **Final Temperature ($T_f$)** | **Change in Temperature ($\Delta T$)** |
| | | |

| Trial 3 (Optional) | | |
|---|---|---|
| **Initial Mass ($m_i$)** | **Final Mass ($m_f$)** | **Change in Mass ($\Delta m$)** |
| | | |
| **Initial Temperature ($T_i$)** | **Final Temperature ($T_f$)** | **Change in Temperature ($\Delta \underline{T}$)** |
| | | |

**Average $\Delta m$:** _____

**Average $\Delta T$:** _____

**Question 1:** As stated in the introduction, the heat energy absorbed by the water in the calorimeter can be calculated using the equation below:

**energy absorbed by water (cal) = mass of water (g) × Δ$T$ (°C) × 1 cal/g °C**

Calculate the heat energy in calories. Show the steps of your calculation, and write your final answer on the line below.

**Remember that 1.00 mL of water has a mass of 1.00 g.**

**heat energy absorbed by water = _____ calories**

**Question 2:** Let's assume (as an exercise) that **all** of the heat released by burning the food was transferred to heating the water. You know how much heat was absorbed by water, and you know the mass of food that was burned (Δ$m$).

Calculate an estimate for the **Calorie content of popcorn** in **Calories/gram** (these are dietary calories). Show the steps of your calculation, and write your final answer on the line below.

**Remember that 1 Calorie = 1000 calories**

**estimated Calorie content of popcorn = _____ Calories/gram**

**Question 3:** All foods are required to list their Calorie content on a nutrition label, which is typically listed as in terms of **serving size**. Use the nutrition label from the bag of popcorn to calculate how many Calories are contained within **1 gram of popcorn**.

Show the steps of your calculation, and write your final answer on the line below.

**1 serving size = 40 grams = 249 kernels = 120 Calories**

**label Calorie content of popcorn = _____ Calories/gram**

**Question 4:** You will find there is a difference between your estimated Calorie content (from Question 2) and the nutritional data obtained from the food label.

Use the equation below to calculate the **% difference** between your estimated value and the label value for the calorie content of popcorn:

$$\% \text{ difference} = \frac{(\text{label value} - \text{estimated value})}{\text{label value}} \times 100$$

% difference =

**Question 5:** The % difference in Question 4 reflects the imperfections of the tabletop calorimeter that you used for this experiment. In the space below, suggest **two modifications to this experiment** (equipment or procedures) that would **improve the accuracy** of your estimated value for the calorie content of popcorn.

1.

2.

# PART B: MEASURE THE CALORIE CONTENT OF ANOTHER FOOD ITEM

**Food item selected:** _____

## Data Table 2: Mass and Temperature Changes for Another Food Item

| Trial 1 | | |
|---|---|---|
| **Initial Mass ($m_i$)** | **Final Mass ($m_f$)** | **Change in Mass ($\Delta m$)** |
| | | |
| **Initial Temperature ($T_i$)** | **Final Temperature ($T_f$)** | **Change in Temperature ($\Delta T$)** |
| | | |

| Trial 2 | | |
|---|---|---|
| **Initial Mass ($m_i$)** | **Final ass ($m_f$)** | **Change in Mass ($\Delta m$)** |
| | | |
| **Initial Temperature ($T_i$)** | **Final Temperature ($T_f$)** | **Change in Temperature ($\Delta T$)** |
| | | |

| Trial 3 (Optional) | | |
|---|---|---|
| **Initial Mass ($m_i$)** | **Final Mass ($m_f$)** | **Change in Mass ($\Delta m$)** |
| | | |
| **Initial Temperature ($T_i$)** | **Final Temperature ($T_f$)** | **Change in Temperature ($\Delta T$)** |
| | | |

**Average $\Delta m$:** _____

**Average $\Delta T$:** _____

**Question 6:** The heat energy absorbed by the water in the calorimeter can be calculated using the following equation:

**energy absorbed by water (cal) = mass of water (g) $\times \Delta T$ (°C) $\times$ 1 cal/g °C**

Calculate the heat energy in calories. Show the steps of your calculation, and write your final answer on the line below.

**heat energy absorbed by water = _____ calories**

**Question 7:** Calculate an estimate for the **Calorie content** of your food item in **Calories/gram** (these are dietary calories). Show the steps of your calculation, and write your final answer on the line below.

**estimated Calorie content of food = _____ Calories/gram**

**Question 8:** According to your calculations, does the second food item have a **larger or smaller** Calorie content per gram when compared to popcorn? Circle which one applies based on your data.

**SMALLER**               **LARGER**

**Question 9:** Use the nutrition label from the bag of food you selected to calculate how many Calories are contained within **1 gram of the food.**

Show the steps of your calculation, and write your final answer on the line below.

**label Calorie content of food = _____ Calories/gram**

**Question 10:** You will find there is a difference between your estimated Calorie content (from Question 7) and the nutritional data obtained from the food label.

Use the equation below to calculate the **% difference** between your estimated value and the label value for the calorie content of popcorn:

$$\% \text{ difference} = \frac{(\text{label value} - \text{estimated value})}{\text{label value}} \times 100$$

% difference =

# EXPLORATION 9
# Synthesis of Aspirin

## Overview

In this lab experiment, you will be synthesizing aspirin, a commonly used pain killer. The experiment will introduce you to synthesis, purification, and quality control techniques that are frequently used in the pharmaceutical industry.

## Preparation

You should be familiar with the following concepts and techniques:

- Natural product that serves as the basis for synthesizing aspirin
- Chemical reaction for synthesizing aspirin
- Chemical test for the purity of the synthesized aspirin
- Calculation of theoretical yield and percent yield

## Equipment and Supplies

Bring a calculator to this lab session.

**Equipment**
Erlenmeyer flask, 125 mL
Graduated cylinder, 10 mL
Beakers, 250 mL, 500 mL
Additional beaker for ice bath
Glass stirring rod
Ring stand and clamp
Hot plate
Buchner funnel and filter paper
Filter flask, tubing, and gasket or drilled rubber stopper

**Supplies**
Salicylic acid, 2.0 g
Acetic anhydride, 4 mL
95% ethanol, 10 mL
Deionized water
1% Ferric chloride ($FeCl_3$)

# INTRODUCTION

One of the most commonly known pain killers and fever reducers is **aspirin** (acetylsalicylic acid). **Salicylic acid** is a natural product found in the bark of the willow tree and is thought to have been used as a pain reliever since the fifth century B.C.E. However, salicylic acid is bitter and can irritate the stomach. Thus, significant effort was made to create a form of the drug that would have the positive pain-relief effects, but not the negative side effects, of salicylic acid.

The first synthesis of **aspirin** occurred in 1897 and is largely credited to Felix Hoffman, an employee of Friedrich Bayer & Company. His synthesis reacted salicylic acid with acetic anhydride to crease **acetylsalicylic acid**. The commercialization of this product, which was branded as aspirin, began in 1899 and led to Bayer being regarded as the world's first pharmaceutical company.

In this experiment, you will synthesize aspirin by using a method like that of Felix Hoffman. After the synthesis, you will purify your product, determine the percent yield, and test the product for purity. The reaction that you will run is shown in Figure 1. In this reaction, salicylic acid ($C_7H_6O_3$) is treated with an excess of acetic anhydride ($C_4H_6O_3$) in the presence of a catalyst—in this case, phosphoric acid ($H_3PO_4$). After heating, the products acetylsalicylic acid ($C_9H_8O_4$) and acetic acid ($C_2H_4O_2$) are formed. This type of reaction is called an **esterification reaction**.

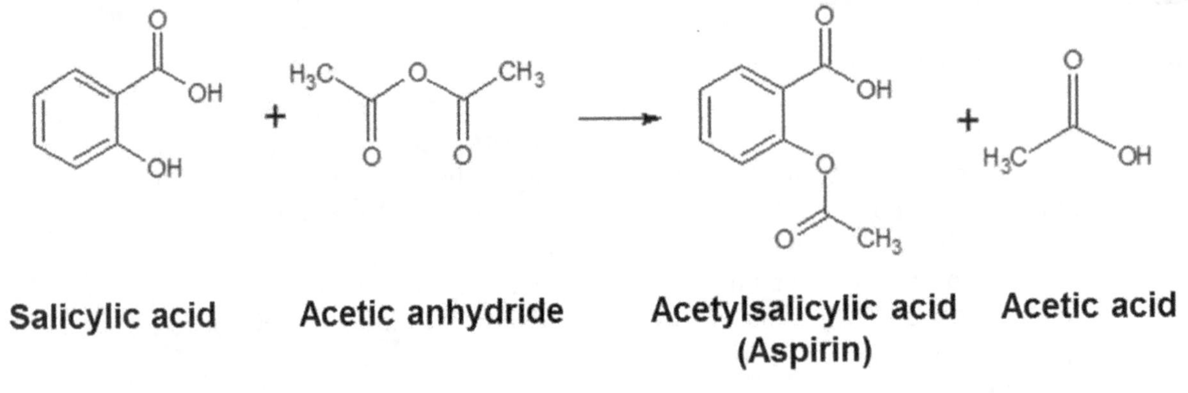

| Salicylic acid | Acetic anhydride | Acetylsalicylic acid (Aspirin) | Acetic acid |

**Figure 1** Synthesis of aspirin from salicylic acid and acetic anhydride.

After the reaction is completed it is necessary to remove any excess anhydride and to crystallize the solid product. Both tasks are accomplished through the addition of water. Most products that are synthesized in the lab contain impurities that must be removed from the final product. In the pharmaceutical industry, obtaining a pure product is very important. For example, commercial aspirin samples can contain 0.15% salicylic acid at most. In this lab, you will perform a recrystallization to further purify the aspirin.

96

In a pharmaceutical laboratory, the purity of the synthesized aspirin is measured with a variety of techniques and instrumentation. However, it is often useful to have a quick purity test that can be performed without the use of expensive instrumentation. One such test that can be used in the synthesis of aspirin utilizes ferric chloride ($FeCl_3$). Ferric chloride is an ionic compound that separates into its ions when added to water. One of the ions, $Fe^{3+}$, will bind tightly with phenols and change color from yellow to purple. A phenol ring, shown in Figure 2, is present in the starting material, salicylic acid, but not in the aspirin product. Thus, the addition of ferric chloride can be used as a test to ensure that no starting material remains in the product.

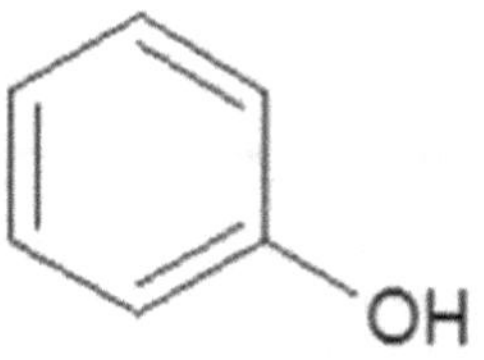

**Figure 2** The molecular structure of a phenol ring. This structure is present in salicylic acid but is not present in aspirin.

---

## WORKED EXAMPLE

**Theoretical Yield of Aspirin**

The theoretical yield of aspirin is the maximum mass of aspirin that could be formed from the reaction of salicylic acid and acetic anhydride. As acetic anhydride is added in excess, this yield depends on the initial amount of salicylic acid used in the reaction. A calculation of theoretical yield also requires the molar mass of the reactant (138.1 g/mol, salicylic acid) and product (180.2 g/mol, aspirin) as well as the mole-to-mole ratio of salicylic acid to aspirin from the balanced chemical equation (1:1 for this reaction). For example, if you begin the reaction with 3.00 g of salicylic acid, the maximum possible yield of aspirin is 3.91 g, as shown in the calculation below.

$$\text{theoretical yield} = 3.00 \text{ g salicylic acid} \times \frac{1 \text{ mol salicylic acid}}{138.1 \text{ g salicylic acid}} \times \frac{1 \text{ mol aspirin}}{1 \text{ mol salicylic acid}} \times \frac{180.2 \text{ g aspirin}}{1 \text{ mol aspirin}}$$

$$= 3.91 \text{ g aspirin}$$

**Percent Yield**

In a chemical reaction, the actual yield of the product is never 100%. Thus, chemists are interested in the efficiency of their reaction. The percent yield of a reaction depends on the theoretical yield and the actual yield of the product formed. For the reaction above, the theoretical yield is 2.61 g. Suppose that 2.33 g of aspirin are formed at the end of the reaction. In this case, the percent yield can be calculated as shown below.

$$\text{percent yield} = \frac{\text{actual yield}}{\text{theoretical yield}} \times 100 = \frac{2.33 \text{ g}}{2.61 \text{ g}} \times 100 = 89.3 \text{ \%}$$

## EXPERIMENTAL PROCEDURES

### SAFETY INFORMATION

<u>CAUTION</u>: Goggles and gloves must be worn for the duration of the experiment.

## Synthesis of Aspirin

1. Weigh out 2.00 g of salicylic acid ($C_7H_6O_6$) onto weighing paper.

2. Transfer the powder to a 125 mL Erlenmeyer flask, and reweigh the weighing paper to determine the actual weight of salicylic acid that is transferred to the flask. Record these numbers in **Data Table 1** on the Data Sheets.

3. In the fume hood, add 4.0 mL of acetic anhydride to the flask.

4. Using a 1 mL disposable pipette, carefully add 6 drops of 85% phosphoric acid to the flask.

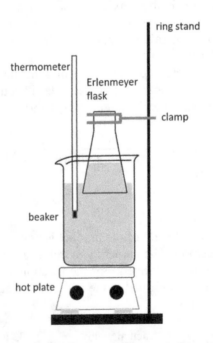

**Figure 3** Setup for a hot water bath with a hot plate.

5.  Add the flask to a water bath as shown in Figure 3. Turn on the hot plate, and heat the water to boiling. Leave the flask in the boiling water for 10 minutes.

6.  Remove the flask from the hot water bath, and set it aside to cool for 5 minutes. After 5 minutes, add 10 mL of deionized water to the flask.

7.  Set up the vacuum filter with a Büchner funnel as shown in Figure 4. Place your filter paper in the funnel, and use a few drops of deionized water to wet the filter. Be sure that all of the holes in the funnel are covered by the filter paper.

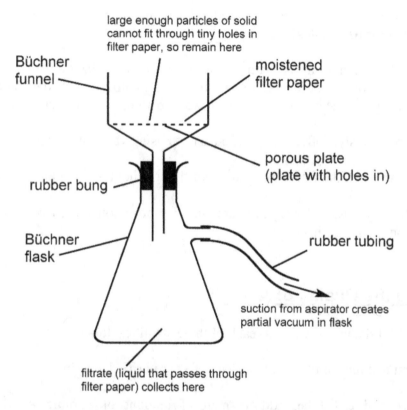

**Figure 4** Vacuum filtration setup with Büchner funnel.

8.  Turn on the vacuum, and transfer your mixture from the flask into the funnel. Rinse the flask with 3 mL of the cold, deionized water to be sure that all the product is transferred to the funnel.

9.  Use the remaining cold water to rinse the funnel.

10. Transfer the product to a watch glass. The filtrate can be discarded down the sink.

## Recrystallizing the Aspirin

1. Save a small portion of your aspirin product in a vial. Label this portion **Sample A**.

2. Transfer the remaining product to a clean, 250 mL beaker.

3. Add 10 mL of 95% ethanol to dissolve the aspirin crystals. If the crystals do not dissolve, place the beaker into a hot water bath, and warm to dissolve the crystals. **DO NOT BOIL** the solution. If necessary, add an additional 2 mL of 95% ethanol.

4. Remove the beaker from the hot water bath, and allow it to cool to room temperature, approximately 10 minutes.

5. Place the beaker into an ice bath to complete the crystallization. If crystals do not appear, scrape the bottom of the beaker to help induce crystallization. If the solution oils rather than crystallizes, place the beaker into the hot water bath until the oil has disappeared.

6. As done previously, collect the purified aspirin using vacuum filtration.

7. Dry the crystals in the funnel by pulling air though them for an additional 10 minutes.

8. Collect the crystals, and weigh the amount of aspirin collected. Record this amount in **Data Table 2** on the Data Sheets.

## Analyzing the Quality of Aspirin

1. Add 5 mL of deionized water to each of three small test tubes.

2. In the first test tube, add a small amount of your purified aspirin.

3. In the second test tube, add a small of amount of Sample A (the aspirin prior to recrystallization).

4. In the third test tube, add a small amount of salicylic acid.

5. Add 10 drops of 1% ferric chloride ($FeCl_3$) to each test tube.

6. Compare the color in the tubes, and record your observations in **Data Table 3** on the Data Sheets.

# CLEAN-UP CHECKLIST

It is essential that you clean up properly after the experiments are complete.

☐ Dispose of the **filtrate from the recrystallization** into the labeled waster container.

☐ Wash all **beakers, flasks, stirrers, and spatulas** with soap and water, and return them to your lab bench.

☐ Dispose of the **ferric chloride** samples into the labeled waste container.

☐ Rinse the test tubes with water, and add the rinse to the ferric chloride waste container. Rinse enough times until the test tube is **CLEAN**. Place **cleaned and empty test tubes** on the **rack** on the lab bench.

☐ Throw all used **weigh boats and paper towels** into the trash.

☐ Clean up any spills on your lab bench.

## DATA SHEETS:
## Synthesis of Aspirin

Name: _____ Date: _____

Lab Instructor: _____ Section: _____

## Synthesis of Aspirin

### Data Table 1: Initial Mass of Salicylic Acid

| Mass of Weigh Paper and Salicylic Acid | Mass of Weigh Paper | Mass of Salicylic Acid Added to Flask |
|---|---|---|
|  |  |  |

### Data Table 2: Yield of Aspirin

| Mass of Weigh Paper and Aspirin | Mass of Weigh Paper | Mass of Aspirin |
|---|---|---|
|  |  |  |

## Analyzing the Quality of Aspirin

### Data Table 3: Ferric Chloride Test

| Sample | Color |
|---|---|
| Purified aspirin |  |
| Sample A |  |
| Salicylic acid |  |

**Question 1:** Assuming the acetic anhydride is added in excess in the reaction, calculate the theoretical yield of the purified aspirin.

*Hint:* See the worked example.

**Question 2:** Calculate the percent yield from the mass of the purified aspirin.

**Question 3:** Your percent yield is most likely not 100%. Give a reason why your yield is not 100%, and explain how this reason specifically affects your results.

**Question 4:** (a) Compare the three colors from the ferric chloride test. Explain what these results tells you about the purified aspirin.

(b) What impurities may be present in your crude aspirin? Where do these impurities go in the recrystallization process?

(c) What is the purpose of the recrystallization process?

# EXPLORATION 10
# Making Synthetic Polymers

## Overview

In Exploration 6, we learned that esters are the products of a chemical reaction between a carboxylic acid and an alcohol. It is possible to repeat this reaction many times to create a molecule called a **polyester**. In fact, a polyester is one example of a class of very large molecules called **polymers**, which are formed by repeatedly linking one or more smaller molecules called **monomers**. In this exploration, you will make three types of **synthetic polymer**. We use this term to refer to non-natural polymers that are synthesized in a chemical laboratory. Synthetic polymers such as polyesters and nylon have widespread uses and regularly impact our daily lives.

## Preparation

You should be familiar with the following concepts and techniques:

- How small molecules react to produce a polymer
- How the connection between monomers changes properties of a polymer
- How similar monomers can yield polymers with different properties

## Equipment and Supplies

**Equipment**
Test tube rack
Mass balance
Wooden sticks

**Supplies**
Test tubes, 10 mL
Phthalic anhydride
Adipoyl chloride
Hexamethylene diamine
Sodium hydroxide
Glycerol

# INTRODUCTION

## Making Polyesters

The general chemical reaction for the formation of esters is provided in Figure 1. This reaction is called a **condensation reaction** because it involves the joining of two molecules to produce a larger molecule, with the release of a small molecule as a product—in this case, $H_2O$. The labels $R_1$ and $R_2$ refer to two carbon-containing groups of atoms. Variability in $R_1$ and $R_2$ enables many different esters to be synthesized from the same reaction.

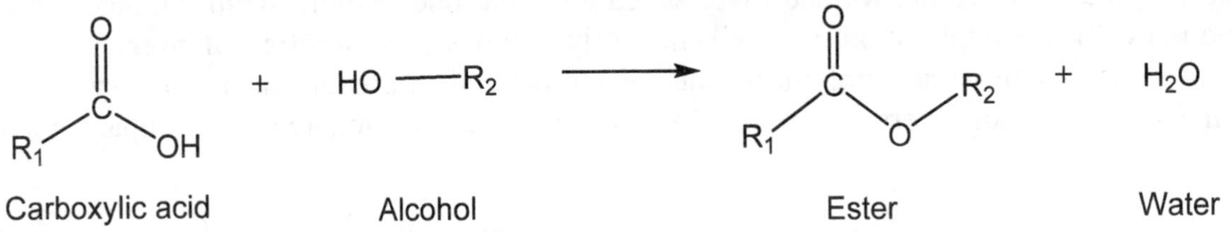

Carboxylic acid                    Alcohol                         Ester              Water

**Figure 1** The chemical reaction for making an ester by a condensation reaction between a carboxylic acid and an alcohol. The labels $R_1$ and $R_2$ refer to two carbon-containing groups of atoms.

In the equation shown in Figure 1, the reactant molecules each have one reactive functional group. As a result, the condensation reaction produces a product containing one ester group, and then the reaction stops. Suppose instead that we use reactant molecules that have **two** reactive functional groups, as shown in Figure 2. **Terephthalic acid** is called a **diacid** to indicate the presence of two carboxylic acid groups. Similarly, **ethylene glycol** is called a **dialcohol**.

The first condensation reaction joins the two monomers to form a product containing an ester group, which is the same as the reaction shown in Figure 1. In this case, however, the product molecule still contains two reactive functional groups. Therefore, it is possible to repeat the condensation reaction many times to create a long molecule that is connected by ester linkages. This process is called **polymerization** and the resulting molecule is called a **polymer**. This molecule is one example of a **condensation polymer** because the monomers are linked by condensation reactions. Because polymers are very long molecules, chemists represent them by using a shorthand that shows the **repeat unit** enclose in square brackets. The repeat unit is the collection of atoms that is repeated multiple times within the polymer, as indicated by the subscript *n* placed next to the square brackets.

The specific polymer shown in Figure 2 is called **polyethylene terephthalate (PET),** which is derived from the names of two monomers. This polymer is sold using the brand names Dacron or Terylene and has wide range of uses that include ropes, tire cord, food packaging, and plastic films. In addition, many soda, juice, and water bottles are produced from PET.

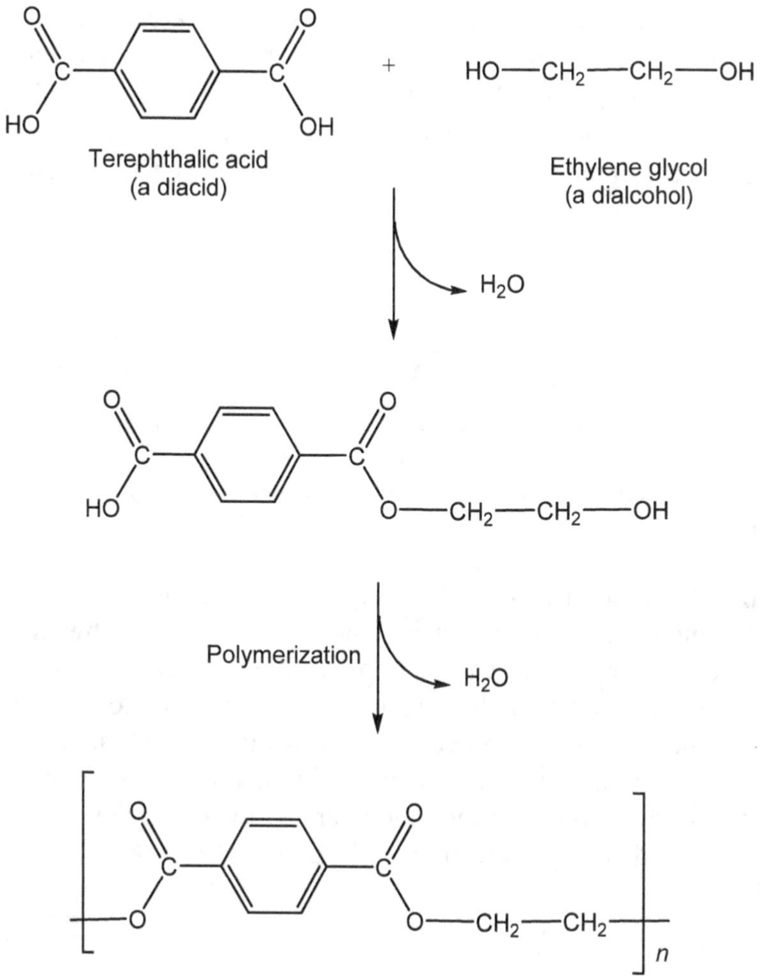

**Figure 2** A condensation reaction between terephthalic acid (a diacid) and ethylene glycol (a dialcohol) produces a product molecule containing an ester group. The remaining functional groups enable the reaction to be repeated many times in a process called polymerization. The result is a polyester molecule with a repeat unit shown in square brackets.

With a choice of dozens of diacids and diols, a very large number of different polyesters can be synthesized. The example shown in Figure 2 is called a **linear polymer** because all of the molecules are linked together in a long, chain-like structure. In a linear polymer, each chain is a separate molecule. However, if either the acid or the alcohol has more than two functional groups, the resulting polymer extends in three dimensions to form a **crosslinked polymer**.

## Making Nylon

A different type of linear polymer called **nylon** was developed at the DuPont company in the 1930s. Nylon is based on an amide linkage, not an ester linkage, between monomers. Amides tend to be more stable than esters, so polyamides have become important materials for textiles and outdoor equipment.

An amide is formed by a condensation reaction between a carboxylic acid and an amine, which releases $H_2O$ as a product. However, the formation of the amide is enhanced by replacing the OH atoms in the carboxylic acid by a Cl atom, which produces a molecule called an **acid chloride** (Figure 3). The condensation reaction between an acid chloride and an amine still generates an amide, but the other product is hydrochloride acid (HCl) instead of $H_2O$.

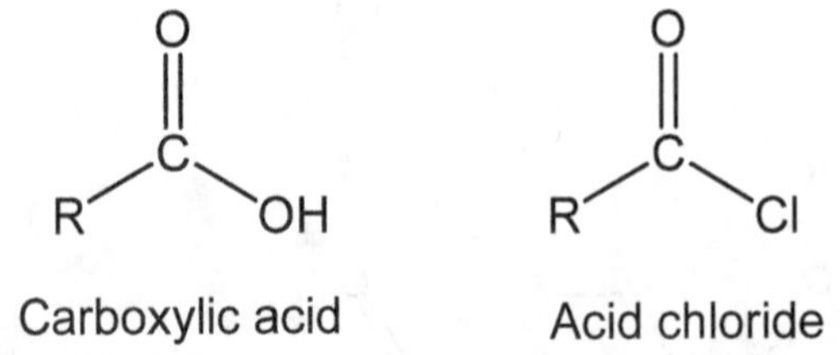

Carboxylic acid          Acid chloride

**Figure 3** The OH atoms in a carboxylic acid can be replaced by a Cl atom to form an acid chloride.

The polymerization reaction that produces nylon is shown in Figure 4. The monomers are **adipoyl chlorine**, a molecule with two acid chloride groups, and **hexamethylene diamine**, which contains two amine groups. In each monomer, the two reactive groups at either end are separated by a saturated hydrocarbon chain containing six carbon atoms. The resulting polymer is called **nylon 6,6** to indicate that the repeat unit (shown in square brackets) contains six carbon atoms from each monomer. In principle, many different types of nylons can be produced by varying the lengths of carbon chains in the monomers. However, nylon 6,6 has proved to be the most useful polymer for industrial applications.

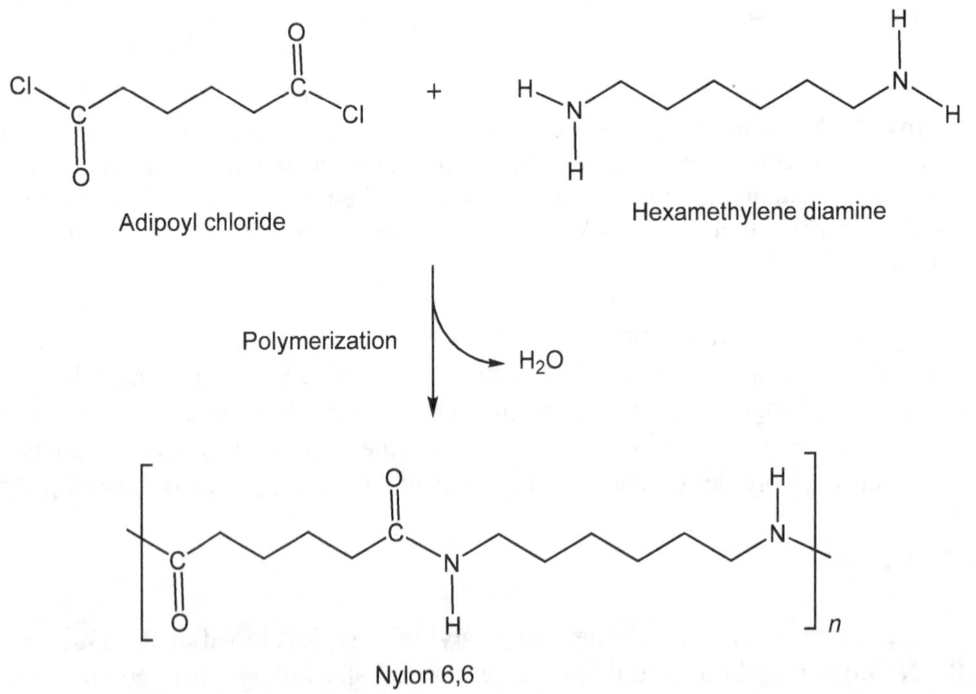

Adipoyl chloride          Hexamethylene diamine

Polymerization          $H_2O$

Nylon 6,6

**Figure 4** Synthesis of nylon 6,6, a condensation polymer, from adipoyl chloride and hexamethylene diamine.

# EXPERIMENTAL PROCEDURES

## Making Linear PET

We begin by synthesizing linear PET. For convenience, we use phthalic anhydride as the starting material for the reaction. When exposed to water, phthalic anhydride is converted into phthalic acid (Figure 5).

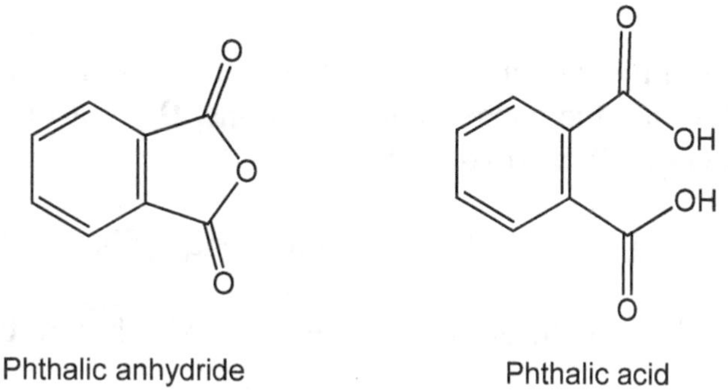

Phthalic anhydride                Phthalic acid

**Figure 5** When exposed to water, phthalic anhydride is converted into phthalic acid (a diacid).

1.  Use a mass balance to obtain 2 g of phthalic anhydride and 0.1 g of sodium acetate.

2.  Add the two substances to a 10mL test tube.

3.  Add 0.8 mL of ethylene glycol to the tube.

4.  Heat the tube very carefully with a burner for 5 to 8 minutes. The sample will appear to boil as water is released during the condensation reactions. **During the heating, keep the tube pointed away from yourself or anyone else.**

5.  Allow the contents to cool until they have the consistency of honey.

6.  Test the thickness using two sticks to determine how readily you can pull a fiber from the mix.

7.  Answer **Question 1** on the Data Sheets.

## Making Crosslinked PET

1.  Use a mass balance to obtain 2 g of phthalic anhydride and 0.1 g of sodium acetate.

2.  Add the two substances to a 10 mL test tube.

3. Add 0.8 mL of glycerol to the text tube. Notes the different structure of glycerol (Figure 6) compared to the structure of ethylene glycol shown in the introduction.

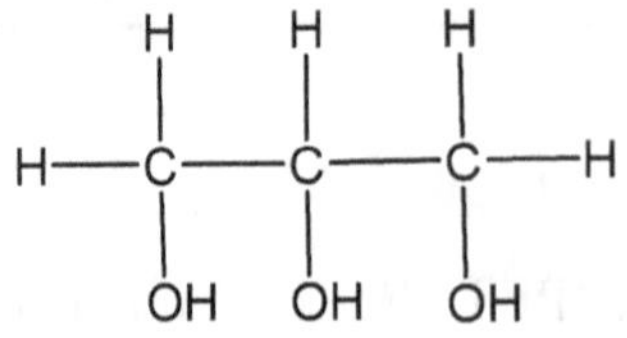

Glycerol

**Figure 6** Molecular structure of glycerol.

4. Heat the tube very carefully with a burner for 5 to 8 minutes. The sample will appear to boil as water is released during the condensation reactions. **During the heating, keep the tube pointed away from yourself or anyone else.**

5. Allow the contents to cool until they have the consistency of honey.

6. Test the thickness using two sticks to determine how readily you can pull a fiber from the mix.

7. Answer **Questions 2 and 3** on the Data Sheets.

## Making Nylon

---

### SAFETY INFORMATION

**CAUTION: This part of the experiment must be performed in a fume hood.**

---

1. Select two 150 mL beakers.

2. To one beaker, add 15 mL of a 5% solution of adipoyl chloride.

3. To the second beaker, add 15 mL of a 5% solution of hexamethylene diamine.

4. Add 10 drops of 20% sodium hydroxide solution to the beaker containing hexamethylene diamine.

5. **Carefully and slowly,** pour the adipoyl chloride solution down the wall of the other beaker. This will form two layers, with a film of polymer at the interface.

6. Using a copper wire of about 15 cm bent into a hook at one end, carefully pull the polymer fibers. and collect the mass at the center of the beaker. Raise the hook slowly to pull a fiber from the polyamide layer until the fiber forms a rope several centimeters long. Dry the rope on a paper towel, and inspect the product.

7. Using tweezers, pull the ends of the rope apart. Feel the resistance. Nylon is spun from the polymer mix to form tight fibers that can be formed into thread. Try twisting the product of your synthesis to see if this leads to a stronger product.

8. Answer **Question 4 to 6** on the Data Sheets.

---

# CLEAN-UP CHECKLIST

---

It is essential that you clean up properly after the experiments are complete.

❑ Empty the test tubes, and dispose of the contents in the appropriate waste containers.

❑ Return all supplies to your bench.

---

# DATA SHEETS:
## Making Synthetic Polymers

---

Name: _____ Date: _____

Lab Instructor: _____ Section: _____

## Making Linear PET

**Question 1:** Record your observations of the consistency of the linear PET polymer.

## Making Crosslinked PET

**Question 2:** Record your observations of the consistency of the crosslinked PET polymer.

**Question 3:** Compare and contrast the properties of the linear and crosslinked PET polymers. Explain how any difference in properties is related to the different molecular structures of the two polymers.

## Making Nylon

**Question 4:** Record your observations of the consistency of the nylon polymer.

## General Questions

**Question 5:** Recycling PET products such as bottles is highly desirable. Which PET product do you think would be easier to recycle, the linear or the crosslinked polymer? Explain your answer.

**Question 6:** Amide bonds are individually stronger than ester bonds. Based on this information, explain why nylon, not PET, is used to make strong ropes for mountain climbing and other purposes.

# EXPLORATION 11
# Digestion of Sugars

## Overview

Digestion is the process of breaking down large molecules into smaller ones, which can then be absorbed by cells in the body and used as nutrients. This disassembly of large food molecules (carbohydrates, proteins, and fats) into their smaller molecular components is achieved by enzymes located in various regions of the digestive system. Each enzyme is optimized to function with maximum efficiency within its particular physiological milieu. For example, an enzyme called amylase exists in saliva and initiates the digestion of starch by cutting it into smaller sugar units, whereas other digestive enzymes in the stomach operate within this very acidic environment. In this exploration, you will investigate the digestion of starch by an enzyme called pancreatic amylase, which converts starch to glucose.

## Preparation

You should be familiar with the following concepts and techniques:

- Sources of starch and its role in our diet
- Classification of sugar molecules
- Role of pancreatic amylase and its physiological environment
- Testing for starch and glucose

## Equipment and Supplies

Bring a calculator to this lab session.

**Equipment**
Finnpipette, 1–5 mL
Ceramic bowl
Hot plate
Forceps
Water bath, 37°C

**Supplies**
Test tubes
Permanent marker
pH paper

**Supplies (continued)**
Disposable pipettes
Parafilm
Conical tubes, 15 mL
Pancreatin α-amylase
Sodium carbonate, 0.5%
Starch, 2%
Hydrochloric acid, 0.2 M
Benedict's reagent
$I_2$/KI solution
Glucose, 2%

# INTRODUCTION

## Starch Digestion

Dietary sources of **starch** include foods such as pasta, rice, bread, and potatoes. Plants synthesize starch as a way of storing energy-rich sugar molecules, linking them together to form a long chain. During human digestion, various enzymes break down large starch molecules into glucose, which can then be used to generate metabolic energy. If excess sugar remains after the body's energy needs are satisfied, the glucose molecules can be linked together again and stored in the liver or muscle as glycogen.

In everyday language, "sugar" usually refers to the sweetener that we add to coffee or what is found in desserts. This molecule is **sucrose**, a disaccharide composed of two sugar units called glucose and fructose. Because they consist of only one sugar unit, glucose and fructose are called **monosaccharides**. Sucrose, with its two sugar units, is called a **disaccharide**. Starch, which contains many sugar units, is a **polysaccharide**. The chains in starch vary in length, but a starch molecule is typically very large, with a molecular mass ranging from several thousand to half a million. This means that several thousand monosaccharides are linked in the chain to form a starch molecule. Figure 1 summarizes the comparison of monosaccharides, disaccharides, and polysaccharides, using a hexagon to represent a single sugar unit. The chemical linkages between sugar units contain an oxygen atom as a bridge between two carbon atoms. Scientists also refer to these various sugar molecules as **carbohydrates**; you are probably familiar with the colloquial term "carbs."

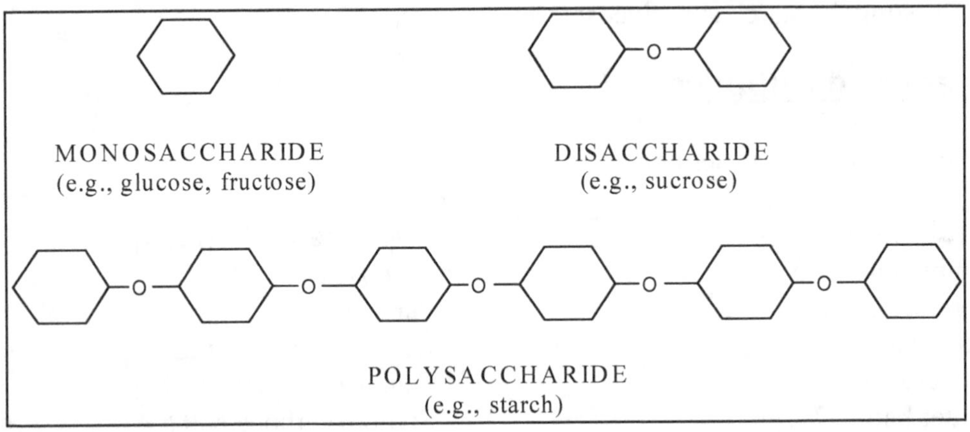

**Figure 1** A schematic comparison of monosaccharides, disaccharides, and polysaccharides. A single sugar unit is represented by a hexagon.

How do humans digest starch? In effect, this process is the reverse of how starch is made from simple sugars. Building a molecule of starch requires removal of a water molecule to create a new chemical linkage between the sugar molecules. The digestion of starch involves adding back the water molecule, thereby breaking the linkage that joins the sugars together. This process is achieved by dedicated enzymes collectively called **amylases**. One such enzyme—**salivary amylase**—functions in the mouth and partially decomposes starch. Many starch molecules escape the action of salivary amylase and continue to pass through the digestive system.

Although we usually associate our stomach with digestion, starch molecules are not processed here. Instead, they travel onto the small intestine, where they are decomposed by an enzyme called **pancreatic amylase,** which is secreted by the pancreas gland. In the human digestive system, pancreatic amylase cuts the long starch molecule into shorter disaccharides, but the final breakdown into monosaccharides is performed by other specialized enzymes. In a test tube, however, pancreatic amylase can function alone to decompose starch into monosaccharides such as glucose.

In this exploration, you will investigate the enzymatic digestion of starch by the pancreatic amylase. You will test for the presence of starch, the starting material, by using $I_2/KI$ solution, and for glucose, the breakdown product, by testing with Benedict's solution.

The iodine test has long been used as a specific test for the presence of starch. The colorless solution will turn dark blue because starch in water adopts a helical configuration and a row of iodine atoms fits neatly into the core of the helix (Figure 2).

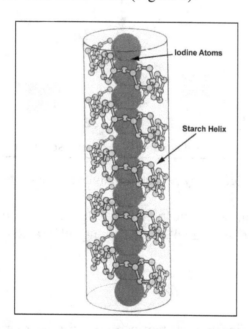

**Figure 2** The helical structure of starch with the row of iodine atoms in the center, which gives rise to the blue color.

The Benedict's test detects the presence of monosaccharides, all of which are reducing sugars. **Starch will not give a positive test with Benedict's solution** since it is a polymer and not a reducing sugar. The Benedict's test solution is blue due to the presence of copper sulfate $(CuSO_4)$. In water, $CuSO_4$ will ionize to copper and sulfate ions; it is the copper ion that reacts with the sugar, producing a red-brown precipitate:

$$CuSO_4 \longrightarrow Cu^{2+} + SO_4^{2-}$$

$$2Cu^{2+} + \text{reducing sugar} \longrightarrow Cu^+$$

$$Cu^+ \longrightarrow Cu_2O \text{ (precipitate)}$$

You will use these two tests in this laboratory to investigate starch digestion. The overall scheme is summarized in Figure 3.

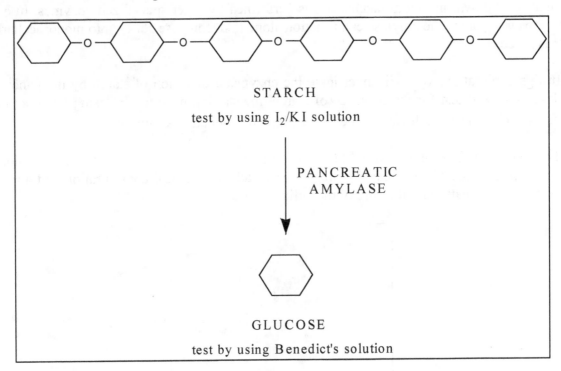

**Figure 3** Starch is digested by pancreatic amylase to generate glucose molecules.

In this lab, you will examine the digestion of starch under different conditions. We will create three different environments by varying pH: acidic (stomach), basic (intestine) and neutral.

## Digestive Disorders

Malfunctions within the digestive system can give rise to a variety of unpleasant disorders. Some conditions arise from a faulty digestive enzyme, whereas some result from infection. The last part of the lab project will involve applying your knowledge of starch digestion to diagnose three individuals who are suffering from various digestive problems. To render the puzzle manageable, we have restricted the possibilities to the four disorders that follow.

### Pancreatic Enzyme Deficiency

The pancreas secretes enzymes that break down fat, carbohydrates, and proteins. People who produce a low level of pancreatic enzyme cannot fully break down food and absorb nutrients.

Symptoms of **pancreatic enzyme deficiency** include abdominal fullness, bloating, and heaviness after eating. Since affected individuals cannot completely digest their food, they often feel hungry and experience sugar cravings very soon after eating a meal. They are often anemic and may show other signs of nutrient deficiency. Treatment of pancreatic enzyme deficiency is accomplished by taking digestive enzyme supplements at the end of each meal. These supplements are derived from either animal pancreatic enzymes or plant enzymes.

## Celiac Disease

**Celiac disease** is an intestinal disorder caused by intolerance to gluten, a protein found in carbohydrates such as wheat, oats, barley, and rye. Ingestion of gluten can cause cramping, bloating, and abdominal pain and results in severe damage to the lining of the small intestine. Celiac disease can be treated with a lifelong gluten-free diet.

## Ulcer

An **ulcer** occurs when acid erosion of the stomach lining provides a suitable environment for growth of a bacterium called *Helicobacter pylori*, which is contracted through food or water. Symptoms include burping, bloating, and nausea as well as severe abdominal pain and burning, which become worse at night and are usually accompanied by vomiting. These symptoms are aggravated by excessive alcohol consumption, cigarette smoking, chocolate, juices, fried foods, and tomato products. Stressful situations can also exacerbate the symptoms, causing the stomach to produce more acid that in turn enhances discomfort. Treatment of ulcers involves taking antibiotics to rid the body of the *H. pylori* bacteria, reducing stomach acid with medication and antacids, and avoiding certain foods that cause indigestion.

## Dysbiosis

**Dysbiosis** occurs when the more than 500 different species of microbial organisms that inhabit the intestinal tract are not balanced in the proper ratio. Affected individuals are not able to properly generate new cells to rebuild the intestinal wall, which usually occurs every 3 to 5 days. Dysbiosis is caused by an altered pH in the intestinal tract, which encourages bacterial overgrowth and hinders digestion and absorption. Symptoms of this disorder include bloating, cramping, headaches, and fatigue. Treatment of dysbiosis incorporates avoidance of drugs that alter pH, such as steroids, antibiotics, antacids, and aspirin, as well as drinking plenty of water, taking vitamin supplements, and increasing fiber and fatty acid intake to nourish and produce new cells.

# EXPERIMENTAL PROCEDURES

## Using the Pipette

The following procedures give you practice with using a **pipette**, which is designed to collect and release a precise volume of liquid.

1.  You will be using a pipette with an adjustable volume (Figure 4). Here are some important points when using the pipette:

    - You can set the pipette to a specific volume in microliters (µL) by **turning the knob at the top**. Please do this **GENTLY** or you will damage the pipette.

    - **ALWAYS ADD A PLASTIC TIP** to the end of the pipette for collecting the sample. If you use the pipette without a tip, you will ruin it! Please return the pipette to the stand when you have finished using it; do **NOT** lay it flat on the bench top.

    - The button at the top of the pipette is used for collecting and dispensing the samples. Pressing the button with your thumb, you will notice that it has **TWO STOPS**. The **FIRST** stop (when your thumb meets some resistance) is used to **COLLECT** the sample. The **SECOND** stop (which requires you to press harder) is used **ONLY** to **DISPENSE** the sample.

**Figure 4**  An adjustable pipette used for small volumes (e.g., 20 µL).

2.  Set the pipette to a volume of **2 mL**. Place a tip on the end of the pipette. **NEVER PLACE THE END OF THE PIPETTE INTO A SOLUTION WITHOUT A TIP!** Push down the knob to the first stop, and place the pipette into the solution to be drawn. Slowly allow the knob to come back up to the top position. You should now have 2 mL of solution in the pipette tip.

3. Place the tip over the 15 mL conical tube, and slowly depress the knob all the way to the second stop to dispense all of the fluid.

4. Practice pipetting the dyed sample into a 15 mL conical tube until you can routinely obtain 2 mL of solution in the tube. Reset the pipette's volume to 1 mL, and practice pipetting the smaller volume into the tube until you can routinely obtain 1 mL of solution in the tube.

## PART A: STARCH DIGESTION

In this lab exercise, you will test the digestion of starch. This involves simulating the conditions of the stomach and the small intestine and testing solutions of different acidities with Benedict's and iodine solutions. First, you will set up the digestion samples, which will incubate for 30 minutes to allow the digestive enzymes to work. During the 30-minute incubation, you will do the controls.

**To accurately test for enzymatic starch degradations, the samples will have to be incubated for 30 minutes at 37°C. You should start with preparing these samples according to Table 1, incubating them, and during the half-hour incubation period, do Part B (Controls).**

1. Use a marker to label three test tubes according to Table 1 below. Follow Table 1 to add the proper solutions to each test tube using the Finnpipette **set to 2 mL**.

*Note:* **Shake the bottle of starch solution before you add it to your test tubes.**

### Table 1: Digestion of Starch Amylase at Different pH Values

| Test Tube | Pancreatic Amylase and Starch Samples | | |
|---|---|---|---|
| **A-1** (Acidic) | 2 mL of pancreatic amylase solution | 2 mL of 0.2 M HCl | 2 mL of starch solution |
| **B-1** (Basic) | 2 mL of pancreatic amylase solution | 2 mL of 0.5% $Na_2CO_3$ | 2 mL of starch solution |
| **C-1** (Neutral) | 2 mL of pancreatic amylase solution | 2 mL of distilled water | 2 mL of starch solution |

2. Now cover each test tube with a small piece of Parafilm. Holding your thumb over the Parafilm, gently vortex each test tube to mix the contents. When you are done, remove the Parafilm.

3. Make sure that your test tubes are **clearly labeled with your name** so that you can identify them. Place all three test tubes (in a test tube rack) in the 37°C (human body temperature) water bath at the front of the room. Set a timer for 30 minutes.

4. Prepare your boiling water bath (to be used for the Benedict's test) by turning on your hot plate and allowing the water in the ceramic bowl to come to a full boil.

5. **Some of the water may evaporate, so you may need to add more water to the bowl. Please note that you will use the light blue test tube rack for your boiling water bath. Do NOT use any of the other plastic racks for the BOILING water bath because they will melt! Leave your boiling water bath on, but remember to refill it because you will use it for the Benedict's test.**

6. **Move on to Part B to test your controls. You will complete the starch digestion procedure after 30 minutes has elapsed.**

## PART B: CONTROLS

To understand the results of this experiment, you will first need to observe what a positive and a negative test for starch and sugar looks like.

1. Set up four test tubes, two for the starch positive and negative controls and two for the sugar positive and negative controls. The negative-control test tubes contain only distilled water. Label them according to Table 2.

2. Follow Table 2 to add the proper solutions to each test tube.

## Table 2: Positive and Negative Controls for Starch and Sugar

| Test Tube | Starch Controls | Procedure for Starch Controls | | | |
|---|---|---|---|---|---|
| 1 | + | **2 mL** of starch | **2 mL** of distilled water | Add **2 drops** of iodine solution to each test tube. Gently swirl each test tube to mix its contents. | Observe your test tubes. Record the results in **Data Table 1** on the Data Sheets. |
| 2 | – | **4 mL** of distilled water | | | |

| Test Tube | Sugar Controls | Procedure for Sugar Controls | | | |
|---|---|---|---|---|---|
| 3 | + | **2 mL** of glucose | **2 mL** of distilled water | Add **15 drops** of Benedict's solution to each test tube. | Place all the test tubes (in the **light blue** test tube rack) in a boiling water bath until a change is observed. | Remove the test tubes and observe. Record the results in **Data Table 1** on the Data Sheets. |
| 4 | – | **4 mL** of distilled water | | | |

# PART A: STARCH DIGESTION (CONTINUED)

1. After 30 minutes have elapsed, retrieve your test tubes from the 37°C water bath.

2. Label three clean test tubes **A-2, B-2, and C-2**.

3. You will now divide the contents of each incubated sample into two equal portions so that you will have sufficient samples for both the starch and the Benedict's test.

4. The original set of tubes **(A-1, B-1, and C-1)** will be used for the Benedict's test and the second set **(A-2, B-2, and C-2)** for the starch test. Follow Tables 3 and 4 to add the proper solutions to your test tubes.

5. Answer Question 1 on the Data Sheets.

**Table 3: Benedict's Test for the Presence of Sugar**

| Test Tube | Samples for Benedict's Test | | | |
|---|---|---|---|---|
| **A-1** | Add ½ of the contents to **A-2**. | Add **15 drops** of Benedict's solution to test tubes **A-1**, **B-1**, and **C-1**. | Place all three test tubes (in the **light blue** test tube rack) in a boiling water bath for 5 minutes. **While you are waiting, perform the starch-iodine test below.** | Remove the test tubes, let cool, and analyze (compare to your positive and negative controls.)Record the results in **Data Table 2** and then answer **Questions 2 through 5** on the Data Sheets. |
| **B-1** | Add ½ of the contents from **B-2**. | | | |
| **C-1** | Add ½ of the contents from **C-2**. | | | |

**Table 4: Iodine Test for the Presence of Starch**

| Test Tube | Samples for Starch Test | |
|---|---|---|
| **A-2** | Add **2 drops** of iodine solution to each test tube. Gently swirl each test tube to mix its contents. | Analyze your test tubes (compare to positive and negative controls), and record the results in **Data Table 2** on the Data Sheets. |
| **B-2** | | |
| **C-2** | | |

# PART C: DIAGNOSING DIGESTIVE DISORDERS

Jen, Tom and Lisa went out for a big pasta dinner. After eating, Jen is experiencing bloating. Since Tom had not eaten pasta in a few months, this meal was a treat for him, but now he is also suffering from abdominal pain and distension. Lisa is bloated and nauseous, feels the onset of a migraine, and is complaining about a severe, cramping pain on the lower left side of her abdomen.

All three are examined by their gastroenterologists. Jen is told she must start taking supplements with plenty of water three times a day. Tom is sent for a consultation with a nutritionist about changing what he eats. Lisa is advised to drink more water, eat more whole grains, and quit smoking. What could be wrong with these individuals? Could they have a digestive enzyme deficiency or some other digestive disorder?

On your lab bench, you have three test tubes containing samples (6 mL each) taken from Jen's, Tom's, and Lisa's small intestines. Your task is to perform the Benedict's and iodine tests to determine which disorders Jen, Tom, and Lisa have. Use the background information on the four digestive disorders provided in the introduction to help you with your diagnoses. **You MUST make your diagnoses based on the tests as well, not just from the information provided in the introduction.** In addition to the Benedict's and iodine tests, be sure to test the pH of the sample by using the pH paper and forceps on your lab bench. Think about the pH of the organ from which the sample was taken.

Record your results in **Data Table 3** and answer **Question 6** on your Data Sheets.

---

# CLEAN-UP CHECKLIST

---

It is essential that you clean up properly after the experiments are complete.

❑ Turn off your boiling water bath.

❑ Discard used transfer pipettes, pipette tips, gloves, conical tubes, and pH paper into the trash.

❑ Pour the Benedict's and iodine waste from your test tubes into the correctly labeled waste containers at the back of the room. **There is a separate waste container for each!** Use the distilled water bottle to rinse the test tubes, and pour the residue into the correctly labeled waste containers as well.

❑ Dump the colored practice solution down the drain.

❑ Place the rinsed test tubes in the broken glass box.

❑ Wipe up any spilled solutions with paper towels.

# DATA SHEETS:
## Digestion of Sugars

Name: _____ Date: _____

Lab Instructor: _____ Section: _____

## PART B: CONTROLS

### Data Table 1: Results of Controls

| Test Tube | Starch Controls | Color Change | Starch Present of Absent? |
|-----------|-----------------|--------------|---------------------------|
| 1 | + | | |
| 2 | − | | |
| Test Tube | Glucose Controls | Color Change | Starch Present of Absent? |
| 3 | + | | |
| 4 | − | | |

## PART A: STARCH DIGESTION

### Data Table 2: Starch Digestion Benedict's and Iodine Test Results

| Test Tube | Color Change | Glucose Present of Absent? | Digestion? (Y or N) |
|---|---|---|---|
| A-1 | | | |
| B-1 | | | |
| C-1 | | | |

| Test Tube | Color Change | Glucose Present of Absent? | Digestion? (Y or N) |
|---|---|---|---|
| A-2 | | | |
| B-2 | | | |
| C-2 | | | |

**Question 1:** Did complete starch digestion occur in the above experiment? Justify your conclusions based on how the iodine test works.

**Question 2:** Is glucose present in test tubes **A-1, B-1, and C-1**? Justify your conclusions based on how the Benedict's test works.

**Question 3:** What was the significance of adding $Na_2CO_3$ to test tube **B-1** and HCl to test tube **A-1**? What does this tell you about the process of starch digestion?

**Question 4:** Do you see any differences between your results in test tubes **C-1 and B-1**? Think about the pH (acidic, basic, or neutral) at which starch digestion is the **most** complete. How does this relate to the physiological environment where starch digestion occurs?

**Question 5:** For each sample, do you see any correlation between the Benedict's test results and the iodine test results? Explain why this should occur.

## PART C: DIAGNOSING DIGESTIVE DISORDERS

### Data Table 3: Benedict's Test and Iodine Test Results
### from Diagnosing Digestive Disorders

| Patient | Benedict's Test Results | Iodine Test Results | pH test results | Diagnosis |
|---------|-------------------------|---------------------|-----------------|-----------|
| Jen | | | | |
| Tom | | | | |
| Lisa | | | | |

**Question 6:** Explain how you came to your conclusions for Jen, Tom, and Lisa. Justify your diagnoses based on your test results and the background information in the introduction.

---

# EXPLORATION 12
## Evaporation of Water, Alcohols, and Other Organic Molecules

---

## Overview

In this exploration, you will examine the relationship between evaporation rates and intermolecular forces.

## Preparation

You should be familiar with the following concepts and techniques:

- Vaporization
- Evaporation
- Intermolecular forces
- Hydrogen bonds

## Equipment and Supplies

Bring a calculator to this lab session.

**Equipment**
2 Thermometers
(more can be used if available)
Filter paper
2 Watch glasses
Rubber bands
Stop watch

**Supplies**
7 Test tubes
Deionized water, 5 mL
Methanol, 5 mL
Ethanol, 5 mL
Butanol, 5 mL
Propanol, 5 mL
Hexanes, 5 mL
Acetone, 5 mL

# INTRODUCTION

Matter typically occurs in one of three phases: solid, liquid, or gas. Which state the matter is in depends on external factors as well as the properties of the molecule(s) in question. For example, at room temperature, water ($H_2O$) is a liquid, while oxygen ($O_2$) is a gas. External factors such as temperature can also play a role in determining the phase of matter. When water is cooled to below 0°C, it will transform into a solid, and when heated to 100°C, it will transform into a vapor. By definition, a **vapor** is a gas that can be condensed into a liquid. When matter goes through a phase transformation from liquid to vapor, it is called **vaporization**.

**Evaporation** is a vaporization process in which a substance transforms from the liquid phase to the gas phase at the surface of the liquid. This process is endothermic, which means the substance that is evaporating gains energy in the form of heat. Since energy is conserved, the surroundings must simultaneously lose energy and cool down. This is the reason you sweat on a hot day or when working out. The water on your skin will evaporate, thus cooling the surface of your skin.

To better understand evaporation, it is necessary to consider the **intermolecular forces** that occur between molecules in a solution. As opposed to **intramolecular forces**, which are the covalent bonds between atoms in a molecule, intermolecular forces occur between neighboring molecules. Perhaps the best example of this is the **hydrogen bonds** exhibited by water molecules (Figure 1). Hydrogen bonds are weak bonds that occur due to an electrical attraction between a negatively charged oxygen atom and a positively charged hydrogen atom. These hydrogen bonds hold water together in both ice and liquid water. To change phase, it is necessary to break these hydrogen bonds. When water undergoes a phase change from liquid to vapor, **all** the hydrogen bonds must be broken.

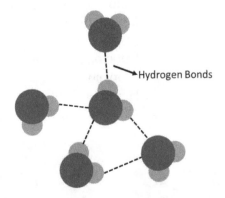

**Figure 1** Hydrogen bonding in liquid water.

Hydrogen bonding can be observed in any molecule where there is a partial charge separation. This is most often observed in molecules that contain a bond between oxygen and hydrogen. A common organic molecule that has hydrogen bonds are **alcohols**. Due to the hydrogen bonding, alcohols are expected to evaporate more slowly than the corresponding alkanes. The chemical structures of the alcohols and alkanes used in this exploration are shown in Figure 2.

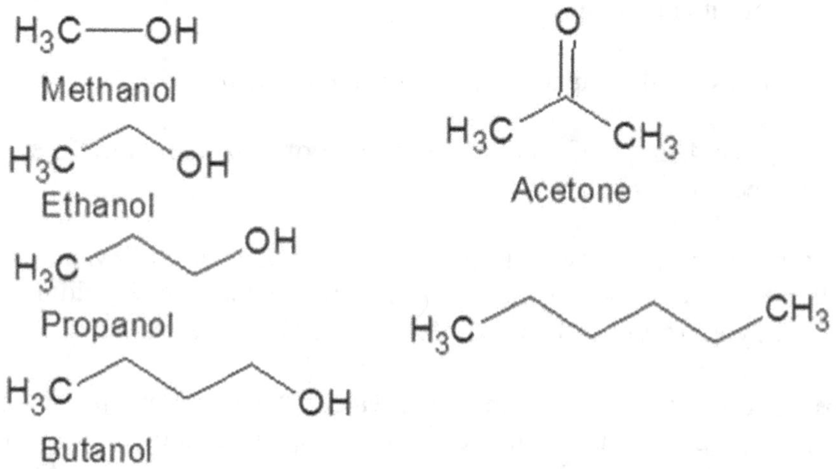

**Figure 2** Line structures of the organic molecules studied in this lab. Note that hexanes typically are sold as a mixture of isomers. Only the structure of *n*-hexane is shown here.

In this exploration, you will explore the rate of evaporation and temperatures changes for water and a number of different organic molecules. Using your data and the molecular structures of the various compounds, you will then explore the relationship between evaporation rate and the temperature change of the surroundings. The effects due to molecular mass and hydrogen bonding will be examined as well.

---

# EXPERIMENTAL PROCEDURES

---

## Evaporation Rates of Compounds

1. Place 2 to 3 mL of the different solvents provided (water, methanol, ethanol, propanol, butanol, hexanes, and acetone) into six separate test tubes. Cap each test tube with a rubber stopper or cork until it is needed.

2. Cut 4 identical pieces of filer paper into 1-inch × 2-inch strips.

3. Using pipettes, place 4 drops of methanol onto one of your pieces of filter paper, and then place the filter paper on a watch glass.

4. Using your stop watch, time how long it takes for the methanol to evaporate from the filter paper. It will be important to start the stop watch immediately after adding the methanol to the paper. Record your results in **Data Table 1** on the Data Sheets.

5. Repeat Steps 3 and 4 for ethanol, butanol, and water. (*Note:* It may take some time for water to completely evaporate. You may move on to the other parts of the experiment while the water evaporates. Just be sure to keep an eye on your water sample to be sure that you record the correct time).

## Thermodynamics of Evaporation

1. Cut 5 more identical strips of filter paper into 1-inch × 2-inch strips.

2. Wrap the end of a thermometer with filter paper, and secure the paper with a rubber band.

3. Lay this thermometer on the lab bench so that the end rests in the air. Be sure to select a location that is free of drafts and where the thermometer will not be disturbed or knocked off the bench top during the course of the experiment. Allow thermometer to rest for 2 to 3 minutes, and then record the temperature on the Data Sheets. This thermometer will be your **Control**.

4. Wrap the end of a second thermometer with a strip of filter paper, again securing the paper with a rubber band.

5. Check to make sure that the second thermometer reads the same temperature as your control. If not, wait 2 to 3 minutes before proceeding.

6. Place the end of the second thermometer into the test tube containing water. Allow the probe to soak in the liquid for approximately 30 seconds.

7. Remove the thermometer, and note the temperature in **Data Table 2**. This temperature is the **initial temperature**, $T_i$.

8. Set the thermometer beside your control thermometer, and wait for 2 minutes.

9. After 2 minutes, note the temperature of the thermometer that had soaked in the water, and record this in **Data Table 2**. This temperature is the **final temperature**, $T_f$.

10. Repeat Steps 5 through 10 for methanol, ethanol, and butanol. Be sure to allow the thermometer to return to room temperature prior to each experiment.

11. Complete the **Data Table 2**, and answer **Questions 1 and 2** on the Data Sheets.

## Making Predictions: Evaporation of Propanol

1. Using your previous results, make a prediction for the evaporation time and temperature change for propanol. Record your predictions in **Data Table 3** on the Data Sheets.

2. Repeat the above experiments for propanol to measure the evaporation rate and temperature change due to evaporation of propanol. Record your results in **Data Table 3**.

3. Answer **Question 3** on the Data Sheets.

## Evaporation of Hexanes and Acetone

1. Repeat the above experiments for hexanes and acetone. Record the evaporation times and temperature changes in **Data Table 4** on the Data Sheets.

2. Answer **Question 4** on the Data Sheets.

## Evaporation of Water

1. Answer **Question 5** on the Data Sheets.

# CLEAN-UP CHECKLIST

It is essential that you clean up properly after the experiments are complete.

❑ Dispose of your solvents in the appropriately labeled waste containers. The water may be poured down the drain.

❑ Rinse the test tubes with water. Rinse enough times until the test tube is **CLEAN**. Place the **cleaned and empty test tubes** on the **rack** on the lab bench.

❑ Throw out all used **filter paper** into the trash.

❑ Wash all **beakers and thermometers** thoroughly.

❑ Clean up any spills on your lab bench.

# DATA SHEETS:
## Evaporation of Water, Alcohols, and Other Organic Molecules

Name: _____ Date: _____

Lab Instructor: _____ Section: _____

## Evaporation Rates of Compounds

### Data Table 1: Evaporation Rates of Water and Alcohol

| Solvent | Evaporation Time (Minutes, seconds) |
|---------|-------------------------------------|
| Methanol | |
| Ethanol | |
| Butanol | |
| Water | |

## Thermodynamics of Evaporation

**Control temperature, $T(^{\circ}C)$:** _____

### Data Table 2: Temperature Change due to Evaporation

| Solvent | $T_i$ (°C) | $T_f$ (°C) | $\Delta T(^{\circ}C)$ |
|---------|------------|------------|------------------------|
| Methanol | | | |
| Ethanol | | | |
| Butanol | | | |
| Water | | | |

**Question 1:** (a) For the alcohols (methanol, ethanol, and butanol), make two graphical plots showing evaporation rate versus number of carbon atoms and $\Delta T$ versus number of carbon atoms.

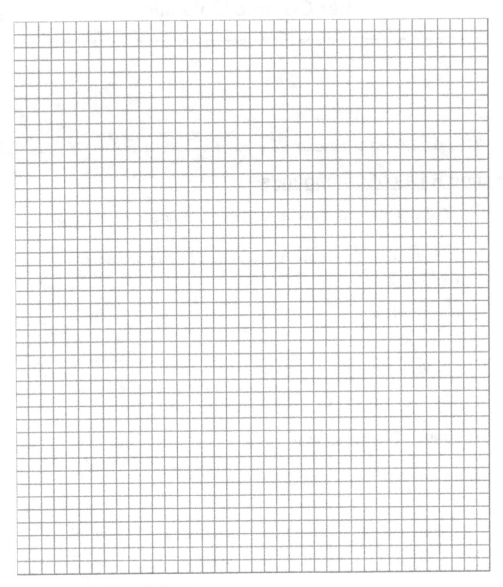

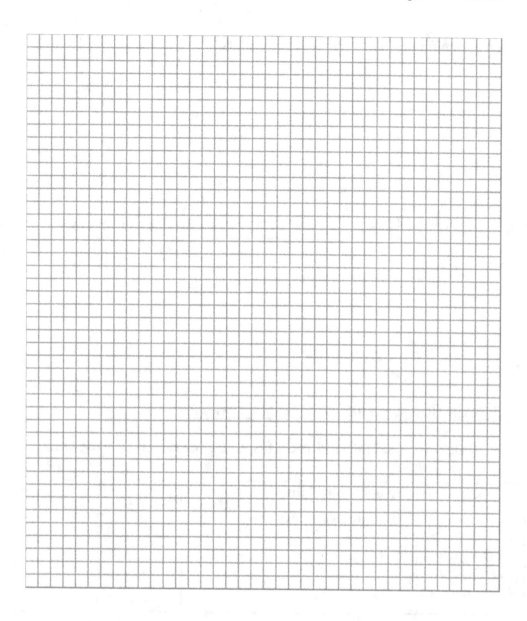

(b) Using the plots, explain the trend between evaporation rate and number of carbon atoms and the trend between $\Delta T$ and number of carbon atoms.

(c) Suggest a reason for the trends observed in the plots.

**Question 2**: Which alcohol has the strongest intramolecular forces, and which has the weakest?

## Making Predictions: Evaporation of Propanol

### Data Table 3: Making Predictions: Evaporation of Propanol

| | Evaporation Time (Minutes, seconds) | $T_i$ (°C) | $T_f$ (°C) | $\Delta T$(°C) |
|---|---|---|---|---|
| Prediction | | | | |
| Measurement | | | | |

**Question 3:** Earlier, you made a prediction for the evaporation rate and $\Delta T$ for propanol. Was your prediction correct? Explain why or why not.

## Evaporation of Hexanes and Acetone

### Data Table 4: Evaporation of Hexanes and Acetone

| Solvent | Evaporation Time (Minutes, seconds) | $T_i$ (°C) | $T_f$ (°C) | $\Delta T$(°C) |
|---------|-------------------------------------|------------|------------|----------------|
| Hexanes |  |  |  |  |
| Acetone |  |  |  |  |

**Question 4:** Structures of a hexane and acetone are shown in Figure 2. You should have noticed that the evaporation rate of hexanes and acetone were faster than the alcohols.

(a) Do hexanes and acetone have stronger or weaker intermolecular forces than the alcohols?

(b) Using the structures provided, explain why hexanes and acetone behave differently than the alcohols.

## Evaporation of Water

**Question 5:** Your water took a significantly longer time to evaporate compared to all the other liquids. Using intermolecular forces, explain why this is the case. Why does water behave so differently?

# EXPLORATION 13
## Solubility of Molecules

## Overview

Vitamins are classified as either **water soluble** or **fat soluble**. The general rule for solubility is the statement that "like dissolves like." This means that polar molecules will dissolve in polar solvents such as water, whereas nonpolar molecules will dissolve in nonpolar solvents such as liquid hexane (which approximates the regions of fat in our body). If a molecule such as methanol ($CH_3OH$) is soluble in water, we can predict that it will not be soluble in liquid hexane.

In this exploration, you will examine the solubility properties of some organic compounds, foods, and vitamins by testing whether they are soluble in water and liquid hexane.

## Preparation

You should be familiar with the following concepts and techniques:

- Difference between intramolecular and intermolecular bonds
- Hydrogen bonding in liquid water
- Dispersion forces/Van der Waals forces
- Test of solubility

## Equipment and Supplies

| **Equipment** | **Supplies** |
|---|---|
| Small test tubes (20 per group) | Cyclohexane |
| Pasteur pipettes with bulbs | Cyclohexene |
| Small spatulas | Octadecane |
| Separatory funnels | Octanol |
| | Ethanol |
| | Glycerol |
| | Olive Oil |
| | Butter |
| | Vitamin C |
| | Vitamin E |
| | Distilled water |
| | Hexane |
| | Food colorings, T Yellow & Green 3 |
| | Triton X-100 detergent solution |
| | Colored pencils |

---

# INTRODUCTION

---

## Solubility of Molecules

The chemical bonds that hold atoms together in molecules are called **intramolecular bonds**. Typically, these intramolecular bonds are strong and require a large input of energy to break. For example, 493 kJ/mol of energy are required to break the covalent bond that joins the O and H atoms in $H_2O$. This bond is not broken by heating water to its boiling point of 100°C.

By contrast, the bonds that connect neighboring molecules are called **intermolecular bonds**. One example of an intermolecular among polar molecules is a **hydrogen bond**. These bonds explain why $H_2O$ is a liquid at room temperature. In liquid water, hydrogen bonds among neighboring $H_2O$ molecules cause the molecules to interact with each other, accounting for water's high boiling point. Based on computer simulations, the typical energy of a hydrogen bond between two $H_2O$ molecules in liquid water is only 8 kJ/mol. This amount is less than 2% of the intramolecular bond energy within $H_2O$. The hydrogen bonds in liquid water can be broken by heating to 100°C, which accounts for the transformation of liquid water into water vapor.

Conversely, nonpolar molecules, such as alkanes, oils, and fats, associate by a different type of interaction called **dispersion forces**. Chemists often refer to these forces as **Van der Waals forces,** after the name of a Dutch scientist. The origin of these forces is a transient, fluctuating dipole interaction between neighboring molecules, which is weak but nevertheless suffices to maintain a nonpolar molecule such as hexane as a liquid.

From the perspective of solubility, these two kinds of interactions account for the difference in behavior of hydrocarbons compared to water, alcohols, or sugars. Methanol is soluble in water in any amount, and the same is true of sucrose (table sugar). These molecules contain one or more polar —OH groups, which can form hydrogen bonds with the $H_2O$ molecules in water. Oil and water do not mix, which is familiar to you from looking at an oil slick in the rain. Alkanes, such as butane or hexane, are not water soluble because their nonpolar character perturbs the interactions among polar water molecules. Instead, these molecules tend to separate themselves when placed in water, and we refer to them as **water-insoluble compounds**.

## A Test of Solubility

In this exploration, you will examine the solubility of various molecules in different solvents. For example, both hexane and hexene are alkanes. Are they equally insoluble in water? Are they equally soluble in octanol? Is ethanol or methanol more soluble in hexane? To answer questions like these, we need an experimental method to measure solubility.

148

Figure 1 illustrates a simple visual test of three possible outcomes that can occur when a solute is mixed with a solvent in a test tube. The first possibility is the formation of a **solution**, in which the solute dissolves completely in the solvent. In a solution, the solvent is distributed evenly throughout; we say that the mixture is **homogeneous**. When a solution is formed, solute is **soluble** in the solvent. The second possibility is the formation of a **suspension**, in which solute molecules clump together to create visible particles. The visual appearance of a suspension is somewhat opaque and "cloudy"; the scientific name for this appearance is **turbid**. In this case, the solute is **insoluble** in the solvent. The third possibility is the formation of a **precipitate**, which is solid matter that collects at the bottom of the test tube. This is another situation in which the solute is insoluble in the solvent. Some substances dissolve rapidly, whereas other are slow. Therefore, each observation must allow sufficient time to reach a valid decision.

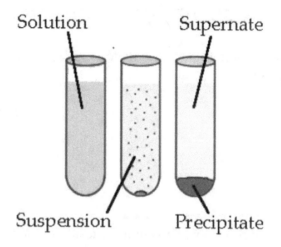

**Figure 1** The visual appearance of a solution, a suspension, and a precipitate.

For the purpose of this exploration, we will classify a solute as either "soluble" or "insoluble" in a particular solvent. However, this binary classification is a simplification of the solubility behavior of molecules. In fact, molecules have a range of solubilities in various solvents. We classify a molecule like methanol as "soluble" in water because it has a high solubility in this solvent. Although we say that methanol is "insoluble" in liquid hexane, a tiny amount of methanol does dissolve in this solvent. So, to be more precise, we can say that methanol has a low solubility in hexane. During your experiments, you may observe cases in which it is not straightforward to classify the solute as soluble or insoluble in a particular solvent. These cases occur when the molecule is partially soluble within a particular solvent.

---

# EXPERIMENTAL PROCEDURES

---

## SAFETY INFORMATION

CAUTION: Goggles and gloves must be worn for the duration of the experiment.

---

## PART A: TEST THE SOLUBILITY OF ORGANIC COMPOUNDS

For the first part of the Exploration, you will test the solubility of the following organic compounds in water and hexane:

Octanol
Octadecane
Glycerol
Cyclohexane
Cyclohexene
Ethanol

1. Label 12 test tubes with the solute and solvent to be tested— for example, **Octanol + Water**.

2. Add approximately 0.5 mL (10 drops) of each liquid solute to the appropriate test tubes.

3. Add approximately 3 to 4 mL of solvent (distilled water or hexane) in small amounts.

4. Cover the tube with a cap, and shake thoroughly until the contents are mixed completely.

5. Repeat steps 2 to 4 for each combination of solute and solvent.

6. If the mixture of solute and solvent forms a **clear solution**, record the result as "soluble" in **Data Table 1** on the Data Sheets.

7. For samples that do not dissolve immediately, continue to shake the tube for **2 minutes** to see if the solute does dissolve over a longer period of time. After shaking, let the sample settle for **2 minutes**.

8. If the mixture of solute and solvent forms a **turbid suspension**, record the result as "insoluble" in **Data Table 1** on the Data Sheets. If you observe that the mixture begins to form two layers after settling, this mixture is also classified as **insoluble**. Record this result in **Data Table 1**.

9. After you have finished the experiments, answer **Questions 1 and 2** on the Data Sheets.

## PART B: EXAMINE THE EFFECT OF A DETERGENT

You will now examine the effect of adding a **detergent** molecule called Triton X-100 on the solubility of the solutes in water.

1. To each tube of a solute in **water**, add 1 drop of Triton X-100 detergent solution.

2. Cover the tube with a cap, and **gently mix** the contents. **Do not shake the tube** because it will create foam.

3. Record your observations in **Data Table 2** on the Data Sheets.

4. Answer **Questions 3 and 4** on the Data Sheets.

## PART C: TEST THE SOLUBILITY OF FOODS AND VITAMINS

In this part of the exploration, you will test the solubility of the following foods and vitamins in water and hexane:

Butter
Olive oil
Vitamin C
Vitamin E

1. Label 8 test tubes with the solute and solvent to be tested—for example, **Butter + Water**.

2. Add each solute to the appropriate test tubes. If the solute is a liquid, add approximately 0.5 mL (10 drops). If the solute is a solid, add approximately 0.1 to 0.2 g (a small spatula tip).

3. Add approximately 3 to 4 mL of solvent (distilled water or hexane) in small amounts.

4. Cover the tube with a cap, and shake thoroughly until the contents are mixed completely.

5. Repeat steps 2 to 4 for each combination of solute and solvent.

6. .If the mixture of solute and solvent forms a **clear solution**, record the result as "soluble" in **Data Table 3** on the Data Sheets.

7. For samples that do not dissolve immediately, continue to shake the tube for **2 minutes** to see if the solute does dissolve over a longer period of time. After shaking, let the sample settle for **2 minutes**.

8. If the mixture of solute and solvent forms a **turbid suspension**, record the result as "insoluble" in **Data Table 3** on the Data Sheets. If you observe that the mixture begins to form two layers after settling, this mixture is also classified as **insoluble**. Record this result in **Data Table 3**.

9. After you have finished the experiments, answer **Questions 5 and 6** on the Data Sheets.

## PART D – PARTITION OF COMPOUNDS BASED ON SOLUBILITY

In the final part of the exploration, you will examine how substances can be separated based on their solubility.

1. Add 10 mL of water and 10 mL of hexane to a 50-mL beaker.

2. To the mixture of water and hexane, add **4 drops each** of the following two food colorings:

   T Yellow
   Green 3

3. Take a **separatory funnel,** shown in Figure 2, and turn the stopcock on the bottom of the funnel to the **closed position**.

4. Pour the contents of the beaker into the top of the separatory funnel. Add a stopper to close the opening at the top.

5. Vigorously shake the separatory funnel for several minutes.

6. Place the funnel in a ring stand, and allow the contents to settle. You will observe the formation of **two layers**, with hexane (less dense) on the top and water (more dense) on the bottom.

7. Answer **Questions 7 and 8** on the Data Sheets.

**Figure 2** A separatory funnel.

# CLEAN-UP CHECKLIST

It is essential that you clean up properly after the experiments are complete.

❏ Dispose of all your solutions in the appropriate waste containers.

❏ Rinse out the separatory funnel and dispose of the mixture in the appropriate waste container.

# DATA SHEETS:
## Solubility of Molecules

Name: _____ Date: _____

Lab Instructor: _____ Section: _____

## PART A: TEST THE SOLUBILITY OF ORGANIC COMPOUNDS

### Data Table 1: Solubility of Organic Compounds in Water and Hexane

| Solute | Octanol | Octadecane | Glycerol | Cyclohexane | Cyclohexene | Ethanol |
|--------|---------|------------|----------|-------------|-------------|---------|
| Water  |         |            |          |             |             |         |
| Hexane |         |            |          |             |             |         |

**Question 1:** Were any of the solutes difficult to classify as either soluble or insoluble in water or hexane? If so, identify the solute, and explain why the classification was difficult.

**Question 2:** Water is a polar solvent, and hexane is a nonpolar solvent. Based on the principle of "like dissolves like," classify the solute molecules as polar or nonpolar. Organize the solutes in two columns as shown below. If you are unsure about the classification, add a question mark (?) next to the name of the solute.

**Polar Solutes**          **Nonpolar Solutes**

## PART B: EXAMINE THE EFFECT OF A DETERGENT

### Data Table 2 :Effect of Adding a Detergent

| Solute | Octanol | Octadecane | Glycerol | Cyclohexane | Cyclohexene | Ethanol |
|---|---|---|---|---|---|---|
| Observations after adding detergent | | | | | | |

**Question 3:** For which solute/solvent mixtures did you observe a **visible change** after adding the detergent? For which mixtures did you **not** observe a s visible change?

**Question 4:** What household items contain detergent molecules? Based on your observations in Data Table 2, what is the purpose of using these detergents?

# PART C: TEST THE SOLUBILITY OF FOODS AND VITAMINS

## Data Table 3: Solubility of Foods and Vitamins in Water and Hexane

| Solute | Butter | Olive Oil | Vitamin C | Vitamin E |
|--------|--------|-----------|-----------|-----------|
| Water  |        |           |           |           |
| Hexane |        |           |           |           |

**Question 5:** Were any of the solutes difficult to classify as either soluble or insoluble in water or hexane? If so, identify the solute, and explain why the classification was difficult.

**Question 6:** Liquid hexane approximates the nonpolar regions of the body where fat molecules are stored. Using this information, classify each of the foods and vitamins as "water soluble" or "fat soluble." Organize the solutes in two columns as shown below. If you are unsure about the classification, add a question mark (?) next to the name of the solute.

**Water Soluble**          **Fat Soluble**

## PART D: PARTITION OF COMPOUNDS BASED ON SOLUBILITY

**Question 7:** In the space below, sketch your observation of the separatory funnel after the two layers have formed. Use colored pencils to indicate the colors of each layer in the funnel.

**Question 8:** Explain your observations in **Question 7** based on the solubility properties of the two food colorings.

# EXPLORATION 14
## Reactions of Aqueous Ions

## Overview

In this lab exercise, you will use visual observations to determine if a chemical reaction took place on mixing solutions and solids. You will then write balanced chemical equations for the observed reactions. Finally, you will identify an unknown water contaminant by observing its chemical reactions.

## Preparation

You should be familiar with the following concepts and techniques:

- Dissociation of ions when in an aqueous solution
- Signs of a chemical reaction
- Solubility rules for aqueous ions
- Writing balanced equations

## Equipment and Supplies

**Equipment**
Spatula
Periodic table

**Supplies**
Test tubes
Permanent marker
Disposable pipettes
Hydrochloric acid, 0.2 M
Copper sulfate, 0.1 M
Lead nitrate, 0.1 M
Barium chloride, 0.5 M
Sodium bicarbonate
Sodium hydroxide, 0.1 M
Aluminum sulfate, 0.1 M
Zinc nitrate, 0.1 M
Sodium sulfate, 0.1 M
Potassium iodide, 0.1 M
Ammonia, 1 M
Sodium chloride, 0.1 M

# INTRODUCTION

## Changes in Matter

In all chemical reactions, bonds between atoms or ions are broken, and new bonds are formed. The most commonly encountered chemical reactions include precipitation, acid-base, and oxidation-reduction reactions (which will be discussed in a later lab). In the reaction below, the ionic bonds of the reactants are broken to form the bonds of the products.

$$CaCO_3(s) + H_2SO_4(aq) \longrightarrow CaSO_4(aq) + CO_2(g) + H_2O(l)$$

Chemical reactions are often, but not always, accompanied by visible changes. In the reaction above, a gas is produced in the form of carbon dioxide ($CO_2$). These observable changes provide evidence that a chemical reaction has occurred.

## Signs of a Chemical Reaction

Signs that a chemical reaction has occurred are the following:

1. **Formation of a Precipitate**
   A **precipitate** is a solid that is insoluble in the reaction mixture. It often settles to the bottom of the container, or it may form a suspension.

2. **Color Change**
   The interactions of some compounds with light make them appear colored, so the **color** of a substance can change when its molecular composition changes as a result of a chemical reaction.

3. **Evolution of a Gas**
   When the reaction product is a **gas** that is insoluble in the reaction mixture, it will bubble out of the mixture. The gas may or may not have a detectable odor.

4. **Temperature Change**
   The majority of chemical reactions involve a net energy change, which is released as heat. This change may cause an increase or decrease of the reaction mixture **temperature** that is detectable by touching the test tube. However, some changes will be too small to detect in this manner.

## Ions in Aqueous Solution

In this lab session, you will be observing chemical reactions in aqueous solutions. Some of the compounds will be ionic compounds that, if soluble in water, are completely dissociated into their corresponding cations and anions. For the dissolution of sodium chloride (NaCl), we can write the following equation, where *(s)* refers to the solid species and *(aq)* indicates an aqueous ion:

$$NaCl(s) \longrightarrow Na^+(aq) + Cl^-(aq)$$

Similarly, for sodium carbonate and calcium chloride:

$$Na_2CO_3(s) \longrightarrow 2\,Na^+(aq) + CO_3^{2-}(aq)$$
$$CaCl_2(s) \longrightarrow Ca^{2+}(aq) + 2\,Cl^-(aq)$$

It is important to realize that compounds such as these are completely ionized in water; there are no ionic compounds of $NaCl$, $Na_2CO_3$, or $CaCl_2$. These types of ionic compounds are often described as strong **electrolytes** because they conduct electricity.

Ions like $Na^+$ or $Cl^-$ are formed from a single atom, and their charge can easily be predicted using the periodic table. In many cases, however, ions contain more than one atom; these are called **polyatomic ions**. One example in the equations shown above is the **carbonate ion, $CO_3^{2-}$**. This ion contains a central carbon atom that is covalently bonded to three oxygen atoms; it behaves as a functional unit with a net charge of 2−. The carbonate ion does **not** dissociate further in aqueous solution (i.e., it does not **form** oxide ions, $O^{2-}$). In other words, it is a **molecular ion**.

The structures for three types of polyatomic ions are shown in Figure 1. In each case, a central atom (C, N, or S) is covalently bonded to three or four oxygen atoms. Only one structure for each ion is shown, although these ions typically exist in various resonance forms. By convention, the entire polyatomic ion is enclosed in square brackets and the net ionic charge is written on the upper right.

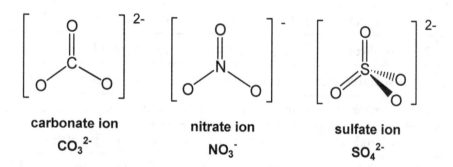

**Figure 1** Structures of three polyatomic ions.

Many metal atoms, such as copper (Cu), iron (Fe), and lead (Pb), can form more than one type of ionic species; for example, copper can form stable ions as $Cu^+$ or $Cu^{2+}$. If the metal ion is part of an ionic compound, its charge can be deduced by examining its ionic partner (Figure 2). For example, lead nitrate has the formula $Pb(NO_3)_2$ and the nitrate ion is $NO_3^-$, so the charge on the lead ion must be $Pb^{2+}$. The charge on the Cu ion in $CuSO_4$ can be deduced in a similar manner.

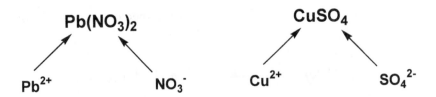

**Figure 2** The charge on a metal ion in an ionic compound can be deduced from the compound's chemical formula.

## Solubility Rules

When aqueous solutions of two different ionic compounds are mixed, sometimes an insoluble solid precipitate separates out of solution. The precipitate that forms is itself ionic; the cation comes from one compound and the anion from the other. To predict the occurrence of reactions of this type, chemists have developed **solubility rules** that categorize which ionic substances are insoluble in water.

Ionic solids cover an enormous range of solubilities. At one extreme is lithium chlorate ($LiClO_3$), which dissolves to the extent of 3.2 kilogram per liter of water (35 mol/L). At the opposite extreme is mercury(II) sulfide (HgS); its solubility is only $10^{-26}$ mol/L. This means that 200 liters of an HgS solution would be required to generate a single pair of $Hg^{2+}$ and $S^{2-}$ ions. The general solubility rules for ionic compounds in water are shown in Table 1.

**Table 1: Solubility Rules for Aqueous Ions**

| $Na^+, K^+, NH_4^+$ | All **sodium, potassium, and ammonium ($NH_4^+$)** compounds are soluble. |
|---|---|
| $NO_3^-$ | All **nitrates** are soluble. |
| $Cl^-$ | Most **chlorides** are soluble, *except* silver, mercury, and lead chlorides. |
| $I^-$ | Most **iodides** are soluble, *except* silver, mercury, and lead iodides. |
| $SO_4^-$ | Most **sulfates** are soluble, *except* strontium, barium, and lead sulfate. |
| $CO_3^{2-}$ | Most **carbonates** are insoluble, *except* those of Group 1 elements and $NH_4^+$ |
| $OH^-$ | All **hydroxides** are insoluble, *except* those of Group 1 elements. |
| $S^{2-}$ | Most **sulfides** are insoluble, *except* those of Group 1 elements and $NH_4^+$. |

When using the solubility rules, the first step is to identify what ions are produced when compounds are dissolved in aqueous solution. As discussed above, polyatomic ions retain their identity as functional units with a net charge. The next step is to examine what combination(s) of ions can produce an insoluble compound, which will be observed as a solid precipitate.

# Writing Balanced Equations

An important component of this lab project will be writing balanced equations for the reactions of ionic compounds in aqueous solution. The two worked examples below illustrate the method of writing chemical equations for an acid-base reaction and a precipitation reaction.

---

### Worked Example 1: Acid-Base Reaction

**Question:** Mixing aqueous solutions of HCl (an acid) and NaOH (a base) produces a chemical reaction. Write a balanced equation for this reaction.

**Answer:** The following species will be present in solution:

$$H^+ \quad Cl^- \quad Na^+ \quad OH^-$$

The $H^+$ from the acid (HCl) is transferred to the base (NaOH) to produce water ($H_2O$) and salt (NaCl). This is a double replacement reaction.

The balanced chemical equation is as follows:

$$HCl(aq) + NaOH\ (aq) \rightarrow H_2O(l) + NaCl(aq)$$

There would be no visible sign of a chemical change in this example, but the pH of the resulting solution (a measure of acidity) will change. If equal amounts of acid and base are added, the resulting solution will have a neutral pH.

---

### Worked Example 2: Precipitation Reaction

**Question:** Mixing aqueous solutions of $Na_2CO_3$ and $CaCl_2$ produces an insoluble precipitate. Write a balanced equation for the chemical reaction, and identify the precipitate.

**Answer:** Since $Na_2CO_3$ and $CaCl_2$ are both soluble, as denoted by the (*aq*), the following ions will float in solution:

$$2Na^+(aq) \quad CO_3^{2-}(aq) \quad Ca^{2+}(aq) \quad 2\,Cl^-(aq)$$

These ions can combine to form NaCl and $CaCO_3$ by "exchanging partners" through the following reaction:

$$2\,Na^+(aq) + CO_3^{2-}(aq) + Ca^{2+}(aq) + 2\,Cl^-(aq) \rightarrow 2\,Na^+(aq) + 2\,Cl^-(aq) + CaCO_3(s)$$

The balanced chemical equation is a double replacement reaction is as follows:

$$Na_2CO_3(aq) + CaCl_2(aq) \rightarrow 2NaCl(aq) + CaCO_3(s)$$

Sodium chloride is soluble, but Table 1 informs us that calcium carbonate is not. Therefore, when these solutions are mixed, an insoluble precipitate of $CaCO_3$ forms and is denoted by (*s*).

---

## EXPERIMENTAL PROCEDURES

## PART A: OBSERVING SIGNS OF CHEMICAL REACTIONS

1.  You will use a disposable pipette to dispense 2 mL of various solutions. Your lab instructor will show you how to use this pipette properly. **You must use a NEW DISPOSABLE PIPETTE every time to avoid contamination of your solutions.**

2.  Label five test tubes with numbers **1 to 5**, and place them in the test tube rack.

3.  Refer to Table 2, and add the chemicals indicated to each of the test tubes in turn. Record the appearance of the reactants before proceeding by using the appropriate box for each reaction provided in **Data Tables 1 to 4** on the Data Sheets.

4.  After the chemicals are mixed, observe each of the test tubes, and record your observations in **Data Tables 1 to 4** for the appropriate box for each reaction.

5.  Answer **Question 1** in the Data Sheets. *Note:* **The equations for the chemical reaction should be written only after all the observations are completed.**

### Table 2: Materials and Amounts for Part A

|   | Reaction | Amounts to be Used |
|---|---|---|
| 1 | $HCl(aq)$ + $NaHCO_3(s)$ | 2 mL + small amount of solid at the end of a spatula |
| 2 | $Pb(NO_3)_2$ $(aq)$ + $BaCl_2(aq)$ | 2 mL + 2 mL |
| 3 | $BaCl_2(aq)$ + $NaCl(aq)$ | 2 mL + 2 mL |
| 4 | $CuSO_4(aq)$ + $NaOH(aq)$ | 2 mL + 2 mL |
| 5 | Solution 4 + $NH_3(aq)$ | 2 mL of $NH_3$ |

# PART B: INDENTIFICATION OF AN UNKNOWN

## Reference Reactions

1. In this experiment, you will mix 0.5 mL of the reactants according to Table 3. Using the plastic pipettes located at your table, fill to the appropriate gradient marking, and dispose after each use. **DO NOT REUSE PIPETTES!**

### Table 3: Materials and Amounts for Reference Reactions

| | **D.** $BaCl_2$ | **E.** $Na_2SO_4$ | **F.** KI |
|---|---|---|---|
| **A.** $Al_2(SO_4)_3$ | 0.5 mL of $Al_2(SO_4)_3$ <br> + <br> 0.5 mL of $BaCl_2$ | 0.5 mL of $Al_2(SO_4)_3$ <br> + <br> 0.5 mL $Na_2SO_4$ | 0.5 mL of $Al_2(SO_4)_3$ <br> + <br> 0.5 mL of KI |
| **B.** $Zn(NO_3)_2$ | 0.5 mL of $Zn(NO_3)_2$ <br> + <br> 0.5 mL of $BaCl_2$ | 0.5 mL of $Zn(NO_3)_2$ <br> + <br> 0.5 mL of $Na_2SO_4$ | 0.5 mL of $Zn(NO_3)_2$ <br> + <br> 0.5 mL of KI |
| **C.** $Pb(NO_3)_2$ | 0.5 mL of $Pb(NO_3)_2$ <br> + <br> 0.5 mL of $BaCl_2$ | 0.5 mL of $Pb(NO_3)_2$ <br> + <br> 0.5 mL of $Na_2SO_4$ | 0.5 mL of $Pb(NO_3)_2$ <br> + <br> 0.5 mL of KI |

2. You will mix each compound in row 1 of Table 3 with each compound in column 1 of the table (i.e., **A + D**, **A + E**, etc.). Label 9 clean test tubes with the appropriate reaction according to the table.

3. Add 0.5 mL of each compound indicated to each test tube, and record your observations in **Data Table 5** of the Data Sheets. You can use the shorthand "**ppt**" to indicate the formation of a precipitate.

4. **After you have completed the table, show your results to your lab instructor before proceeding to the identification of your unknown.**

## Determining an Unknown Contaminant in Water

Based on these observations with known samples, you will now deduce the chemical composition of an unknown sample. Consider a scenario in which you suspect that a source of drinking water from a local well has been contaminated. The suspected pollutants could be aluminum, leached from the soil by acid rain, or either zinc or lead, leaked from a nearby battery factory.

Based on your observation of characteristic chemical reactions in the **Reference Reactions** section, it is possible to obtain information about the substance in the well water.

**For this exercise, your unknown is one of the following:**

$$Al_2(SO_4)_3 \qquad Zn(NO_3)_2 \qquad Pb(NO_3)_2$$

1. Record the unknown that you will be using in the Data Sheets.

2. Study the results that you have generated for the known solutions in **Data Table 5**.

3. Answer **Question 2** on the Data Sheets.

4. You will mix your unknown with the compounds in each column of Table 4. Label 3 clean test tubes with the appropriate reaction according to the table.

5. Add 0.5 mL of each compound indicated to each test tube, and record your observations in **Data Table 6** on the Data Sheets. You can use the shorthand "**ppt**" to indicate the formation of a precipitate.

6. Answer **Question 3** on the Data Sheets.

### Table 4: Materials and Amounts for Unknown Contaminant

|  | **D.  $BaCl_2$** | **E.  $Na_2SO_4$** | **F.  KI** |
|---|---|---|---|
| **Unknown** | 0.5 mL of Unknown<br>+<br>0.5 mL of $BaCl_2$ | 0.5 mL of Unknown<br>+<br>0.5 mL of $Na_2SO_4$ | 0.5 mL of Unknown<br>+<br>0.5 mL of KI |

# CLEAN-UP CHECKLIST

It is essential that you clean up properly after the experiments are complete.

❑ Dispose of the test tubes in the **appropriately** labeled waste container. Rinse the test tube with water, and add the rinse to the waste container. Rinse enough times until the test tube is clean, and place it on the test tube rack on the back bench.

❑ Clean up any spills on your lab bench. Arrange the glassware and solutions neatly before leaving.

# DATA SHEETS:
## Reactions of Aqueous Ions

Name: _____ Date: _____

Lab Instructor: _____ Section: _____

## PART A: OBSERVING SIGNS OF CHEMICAL REACTIONS

### Data Table 1: Observations of Reaction 1

| Reactants | Appearance of Reactants | Appearance of Mixture |
|---|---|---|
| $NaHCO_3$ | | |
| HCl | | |
| **Evidence for a Chemical Reaction:** | | |
| **Balanced Equation:** | | |

### Data Table 2: Observations of Reaction 2

| Reactants | Appearance of Reactants | Appearance of Mixture |
|---|---|---|
| $Pb(NO_3)_2$ | | |
| $BaCl_2$ | | |
| **Evidence for a Chemical Reaction:** | | |
| **Balanced Equation:** | | |

### Data Table 3: Observations of Reaction 4

| Reactants | Appearance of Reactants | Appearance of Mixture |
|---|---|---|
| $BaCl_2$ | | |
| NaCl | | |
| **Evidence for a Chemical Reaction:** | | |
| **Balanced Equation:** | | |

### Data Table 4: Observations of Reactions 4 and 5

| Reactants | Appearance of Reactants | Appearance of Mixture |
|---|---|---|
| $CuSO_4$ | | |
| NaOH | | |
| $NH_3$ | | |
| **Evidence for a Chemical Reaction:** | | |
| **Balanced Equation (for Reaction 4 Only):** | | |

**Question 1:** Write a balanced chemical equation for any reaction you observed in **Data Tables 1 to 4**. Use the periodic table and Table 1 in the introduction to represent the ions with the correct charge.

**In your equations, indicate in parentheses the state of the ion or compound as *(s)* or *(aq)*. If heat is evolved in a reaction (hot tube), write heat as a "product"; if heat is absorbed (cold tube), write heat as "reactant."**

# PART B: INDENTIFICATION OF AN UNKNOWN

## Data Table 5: Observations of Reactions in Part B

|  | **D.** $BaCl_2$ | **E.** $Na_2SO_4$ | **F.** KI |
|---|---|---|---|
| **A.** $Al_2(SO_4)_3$ |  |  |  |
| **B.** $Zn(NO_3)_2$ |  |  |  |
| **C.** $Pb(NO_3)_2$ |  |  |  |

**Label on Unknown:** _____

**Question 2:** Select one or two reactions to try on your unknown. **You should base your conclusion on no more than two reactions**, which is all that is necessary for identification. **Write the reactions that you will use.**

## Data Table 6: Observations of Reactions with the Unknown

|  | D.  $BaCl_2$ | E.  $Na_2SO_4$ | F.  KI |
|---|---|---|---|
| **Unknown** |  |  |  |

**Question 3:** Based on the results from **Data Table 6**, identify your unknown pollutant, and explain your reasoning. Include the balanced equations and identity of the precipitates in your answer.

# EXPLORATION 15
# How Much Vitamin C Is in a Tablet?

## Overview

In this exploration, you will examine a vitamin tablets to determine if its content of vitamin C matches what is claimed on the label. While medicines are packaged in doses that are carefully monitored by the U.S. Food and Drug Administration, supplements such as vitamin C are not monitored by the same agency. We will use a titration experiment to determine the amount of vitamin C in the tablet.

## Preparation

You should be familiar with the following concepts and techniques:

- Symptoms of scurvy as a human disease
- Connection between vitamin C and collagen protein
- Device used to dispense known volumes in a titration experiment
- Chemical reaction used to monitor the amount of ascorbic acid

## Equipment and Supplies

Bring a calculator to this lab session.

**Equipment**
Glass burettes, 50 mL
Burette stands
Volumetric flaks, 100 mL
Volumetric flasks, 200 mL
Pipette, 20 mL
Graduate cylinders, mL
Graduated cylinders, mL
Erlenmeyer flasks, 250 mL

**Supplies**
Samples of vitamin C tablets
$KIO_3$ solution
KI solution
Oleic acid/ethanol solution, 1:200
Starch indicator solutions

# INTRODUCTION

## Vitamin C and Human Health

Why are British people referred to informally as "limeys"? The name has a historical source that is related to chemistry. Sailors on long voyages frequently suffered from a disease called **scurvy**. The early symptoms of scurvy are bleeding of the skin and gums, followed by muscle weakness and overall fatigue that frequently led to death. A month or more without a source of fresh fruit of vegetables is enough to induce symptoms of scurvy. The British navy lost thousands of sailors to the disease. Experienced captains knew that citrus fruits, such as lemons and limes, were effective in counteracting scurvy. Limes in particular were grown in British colonies and were easier to obtain than lemons, which are a better source of vitamin C. So, American slang referred to British sailors as limeys, and the name was eventually extended to everyone from Britain.

The cause of the disease only became clear in the 20th century, after the discovery that humans need several essential vitamins in their diet. Vitamin C is essential for forming skin, bone, and teeth. This vitamin is also called ascorbic acid, which derives from a medical term (*scorbutus*) that was used to describe scurvy. The molecular structure of vitamin C is provided in Figure 1. Many fruits and vegetables are naturally rich in vitamin C, including peppers and broccoli. Meats contain vitamin C as well, but cooking reduces the concentration of the active vitamin. According to the National Institutes of Health, the Recommended Dietary Allowance (RDA) of vitamin C is 90 mg for adult males and 75 mg for adult females.

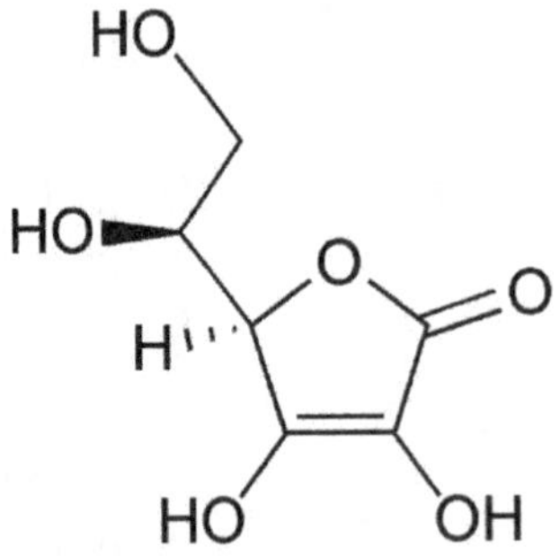

**Figure 1** The molecular structure of vitamin C, also called ascorbic acid.

The biological role of vitamin C is connected to the structure and function of collagen, which is the most prevalent protein in the human body. Approximately 30% of all human protein is collagen and it is found in skin, bone, cartilage, and teeth. As you may expect, the protein required for bone or teeth needs to be rigid and strong. Collagen is unique in that it consists of long, highly repetitive sequences of amino acids, including two that must be chemically modified before they can form bonds to reinforce the protein structure. The chemical changes to the amino acids are essential to stiffen the long chains of collagen.

One of these chemical changes is the conversion of an amino acid called **proline** into a modified version called **hydroxyproline**. A simplified version of the chemical reaction is given below.

$$\text{proline} \ + \ O_2 \ \longrightarrow \ \text{hydroxyproline} \ + \ CO_2$$

This chemical conversion requires an enzyme that functions as a biological catalyst to accelerate the rate of the reaction. The enzyme uses an iron ion, $Fe^{2+}$, to activate the oxygen. To continue its function, the enzyme has to regenerate the iron in its original form after each chemical reaction. The important role of restoring the iron in the enzyme is performed by vitamin C; we say that it acts as a **cofactor** for the enzyme by assisting its function. Without vitamin C, the unmodified amino acid that is inserted into collagen produces an unstable protein. As a result, skin, bone, and teeth made from collagen with unmodified proline lack strength and rigidity and can easily become infected after sufficient time in the body.

## Using Titration to Measure Vitamin C

Many of us take vitamin C tablets as a supplement to ensure a sufficient supply in the body, although a normal diet usually provides enough to achieve the RDA. In this exploration, you will use a **titration** experiment to determine the amount of vitamin C in a commercial tablet. In a titration experiment, we add a known volume of solution from a **burette** (Figure 2).

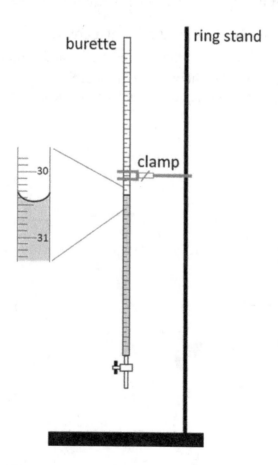

**Figure 2:** Setting up and reading a burette.

Vitamin C is colorless, so we cannot monitor it directly. Instead, we utilize a chemical reaction that involves vitamin C (ascorbic acid). The titration experiment involves adding iodine ($I_2$) to a solution containing ascorbic acid. In practice, we use a mixture of salts (KI and $KIO_3$) because iodine itself is not stable. However, the basic chemistry remains the same. The ascorbic acid reacts with iodine to form dehydroascorbic acid and two iodide ions ($I^-$). This reaction arises from the transfer of two electrons from the ascorbic acid to the iodine.

$$\text{ascorbic acid} + I_2 \longrightarrow \text{dehydroascorbic acid} + 2\, I^-$$

As long as the solution contains some ascorbic acid, it will react with any added iodine. However, the reaction will reach a point at which all of the original ascorbic acid has been converted into dehydroascorbic acid. Because there is no longer any ascorbic acid in the solution, the free iodine can now react with starch in an indicator solution to produce a blue-black color. The presence of this color in the flask indicates the **end point** of the titration.

At the end point, the number of moles of added $I_2$ is equal to the number of moles of ascorbic acid that were originally in the solution.

$$\text{moles of } I_2 = \text{moles of ascorbic acid}$$

We can convert each side of the equation into concentration × volume.

$$\text{concentration of } I_2 \times \text{volume of } I_2 = \text{concentration of ascorbic acid} \times \text{volume of ascorbic acid}$$

From the titration experiment, we know the concentration and volume of $I_2$ solution that was added from the burette. We also know the volume of ascorbic acid solution in the flask. The only unknown is the concentration of ascorbic acid, which can be determined from the equation.

# EXPERIMENTAL PROCEDURES

### SAFETY INFORMATION

<u>CAUTION</u>: **Goggles and gloves must be worn for the duration of the experiment. <u>DO NOT CONSUME</u> any of the tablets in the experiment.**

## Prepare the Vitamin C Sample

**Vitamin C is unstable in the presence of oxygen in air. You must make a fresh sample of the tablet before each experiment.**

1. Select one of the vitamin C tablets provided by the lab instructor. Record the brand name of the tablet and the dose of vitamin C on the Data Sheets.

2. Crush a single tablet using a mortar and pestle.

3. Dissolve the crushed tablet in 200 mL of distilled water. The solution will be slightly cloudy because of the filler in the tablet.

4. Label each sample according to the source of the vitamin.

5. Pipette 20 mL of the sample solution into a 250 mL Erlenmeyer flask, and add 150 mL of distilled water.

6. To the same solution, add 5 mL of the KI solution plus 1 mL of the starch indicator solution.

## Perform the Titration Experiment

You will perform **two trials** of the titration experiment to determine if your result is **reproducible**. If there is a significant difference between the two trials, you will perform a **third trial** to resolve the discrepancy.

1. Fill a burette with the $KIO_3$ solution, which has a concentration of 20 mM. Record the initial volume in **Data Table 1** on the Data Sheets.

2. Add drops of $KIO_3$ solution from the burette until a blue-back color appears in the flask. At first, this color may disappear when you shake the flask.

3. Continue adding drops **slowly** until the blue-black color persists. This is the **end point** of the titration.

4. Record the end point in the **Data Table 1** on the Data Sheets. Calculate the total volume of solution added from the burette.

5. **Repeat the titration experiment** to determine if you reproduce the same result. Record your volumes in **Data Table 1** on the Data Sheets.

6. If the results of your two trials are very close, you can **average** the final volume of the trials. However, if there is a significant difference between your two trials, then you must perform a third trial to resolve the discrepancy.

## Determine the Amount of Vitamin C in the Tablet

You now have the necessary data to determine the amount of vitamin C in the tablet. Follow the procedure provided on the Data Sheets, and answer **Questions 1 through 6.**

---

# CLEAN-UP CHECKLIST

---

It is essential that you clean up properly after the experiments are complete.

❑ Dispose of all solutions in the appropriate waste containers.

❑ Rinse all beakers and the burette with deionized water. The burette should be placed in the holder upside down to allow it to dry.

❑ Wipe down the ring stand and your work area, and discard the paper towels in the trash.

---

# DATA SHEETS:
# How Much Vitamin C Is in a Tablet?

---

Name: _____ Date: _____

Lab Instructor: _____ Section: _____

## Prepare the Vitamin C Sample

Brand name of vitamin C tablet: _____

Dose of vitamin C tablet: _____

## Perform the Titration Experiment

### Data Table 1: Titration of Vitamin C Tablet

| Trial | Initial Burette Reading | Final Burette Reading | Total Volume Added |
|---|---|---|---|
| Trial 1 | | | |
| Trial 2 | | | |
| Trial 3 (If Necessary) | | | |
| Average | | | |

## Determine the Amount of Vitamin C in the Tablet

In the introduction, we learned that the end point of the titration can be expressed as an equation involving the volumes and concentrations of the iodine and ascorbic acid solutions.

concentration of $I_2$ × volume of $I_2$ = concentration of ascorbic acid × volume of ascorbic acid

The equation remains valid provided that we use the **same units** for volume and concentration on each side of the equation. For example, we can use **millimolar (mM)** for concentration and **milliliters (mL)** for volume.

The solution used for the titration is equivalent to an $I_2$ concentration of 60 mM.

177

**Question 1:** (a) Complete the list of the following quantities used in the titration experiment. Use the appropriate unit for each quantity.

Concentration of $I_2$ solution in the burette = _____

Volume of $I_2$ solution added from the burette = _____

Volume of ascorbic acid solution in the flask = _____
*(**Note:** This is the volume of the ascorbic acid solution **before** dilution)*

(b) Use the equation to calculate the concentration of ascorbic acid in the solution. Write your answer in the space below:/

Concentration of ascorbic acid solution in the flask = _____

**Question 2:** Calculate the **number of moles** of ascorbic acid in the flask. Give your answer in **millimoles (mmol)**. Recall that we define a concentration of 1 mM as

$$1 \text{ mM} = \frac{1 \text{ mmol of solute}}{1 \text{ L of solution}}$$

**Question 3:** You made a sample of vitamin C by dissolving the crashed tablet in 200 mL of distilled water. However, the flask in the titration experiment contained only 20 mL of this sample. Calculate how many moles were present in the original 200 mL sample, and give your answer in **millimoles (mmol)**. This number is the total moles of ascorbic acid in the tablet.

**Question 4:** We need to convert the moles of ascorbic acid into a mass to compare with the label. This conversion requires the **molar mass** of ascorbic acid, which has the following chemical formula:

$$C_6H_8O_6$$

Use the following atomic masses to calculate the molar mass of ascorbic acid:

    C  12.0 g/mol             H  1.0 g/mol             O  16.0 g/mol

**Question 5:** Use your answers to **Questions 3 and 4** to calculate the mass of ascorbic acid in the tablet. Give your answer in **milligrams (mg)**.

**Question 6:** (a) You will likely find a difference between your calculated value and the dose reported on the tablet bottle. Use the equation below to calculate the **% difference** between your calculated value and the label value:

$$\% \text{ difference} = \frac{(\text{label value} - \text{calculated value})}{\text{label value}} \times 100$$

% difference =

(b) Did your calculated value closely match the dose of vitamin C reported for the tablet? Was there a significant difference between the two values?

(c) If there was a significant difference between the two values, suggest two possible reasons why your results differ from the vitamin C dose reported on the label.

---

# EXPLORATION 16
## Diffusion and Osmosis

---

## Overview

An average 70 kilogram (154 lb.) adult human consists of approximately 100 trillion cells organized into tissues and organs. Each cell is separated from its extracellular environment by a cell membrane, which regulates the continuous exchanges of chemical substances between the fluid inside the cell (**intracellular fluid**) and the fluid outside the cell (**extracellular fluid**). All living organisms require the transport of biological molecules across membranes. These molecules as well as ions are continually in flux across cell membranes. The cell's transport mechanisms are essential for regulating this flux to satisfy the cell's metabolic requirements. In this exploration, you will observe some of the physical processes involved in the passage of substances across the cell membrane.

## Preparation

You should be familiar with the following concepts and techniques:

- Cell membrane composition
- Definitions for solute, solvent, and solution
- Passive transport (diffusion and facilitated diffusion)
- Osmosis (including the semipermeable membrane and tonicity)
- Active transport

## Equipment and Supplies

**Equipment**
Balance
Beakers, 100 mL, 800 mL, 500 mL
Hot plate
Ceramic bowl
Cork borer
Scalpel
Cork
Hydroponic test tubes
Dialysis tubing
Forceps

**Supplies**
Potato
Weigh boat
Test tubes
Disposable pipettes
Laminated graph paper, 1 $cm^2$ grid
Permanent marker
Sodium chloride, 20% and 0.9%
Starch solution, 0.25% starch in 1% sodium sulfate
Glucose, 10%
Benedict's Reagent
$I_2/KI$ solution

---

# INTRODUCTION

---

## Composition of the Cell Membrane

The **cell membrane** is the gatekeeper for the vital exchanges that take place between the interior of cells cell and their extracellular environment. This membrane is composed of phospholipids, proteins, glycoproteins, glycolipids, and other molecules. These molecules form a semipermeable membrane with important structural, chemical, and electrical characteristics, which include:

1. The **phospholipid** structure favors the penetration of substances that can dissolve in its hydrophobic interior.

2. **Membrane proteins** selectively facilitate the transport of molecules or charged ions, such as negatively charged ions (**anions**) and positively charged ions (**cations**). Each membrane protein typically allows only a specific molecule or ion to pass through it and blocks other chemical substances.

## Transport Processes Across the Cell Membrane

Figure 1 shows the major types of biological transport processes. Substances move in and out of cells by several physical processes described below.

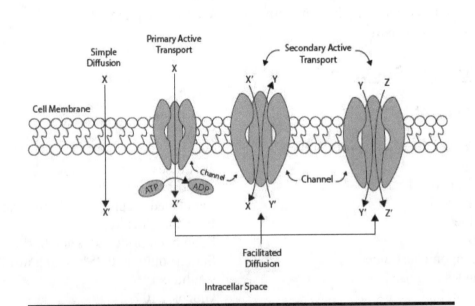

**Figure 1** A cell membrane showing transport processes.

During this exploration, we will commonly refer to the components of a solution. A **solute** is a substance that dissolves in a liquid called the **solvent**, and the resulting mixture is the **solution**. Usually, the solute is present in a smaller amount, whereas the solvent is present in a larger amount. For example, if you had a 20% glucose solution (by mass), you would have 20 g of glucose, the solute, and 80 g of water, the solvent.

## Passive Transport

**Passive transport** involves the spontaneous movement of molecules from a region of higher concentration to a region of lower concentration. We refer to this difference in concentration as a **concentration gradient**. As an analogy, think of the "gradient" of a road, which measures its inclination or slope. Molecules spontaneously move down a concentration gradient just as a ball spontaneously rolls downhill. Passive transport does not require any input of energy to make the molecules move.

## Diffusion

**Simple diffusion** is a form of passive transport in which molecules move from a region of higher concentration to one of lower concentration. The respiratory gases, oxygen and carbon dioxide, pass freely through the plasma membrane. Oxygen passes from its high concentration in oxygenated blood into the cells, where it is continuously used for cell metabolism. Carbon dioxide, on the other hand, is produced by cell metabolism, so it diffuses according to its concentration gradient out of the cell and into the blood.

One important medical application of simple diffusion is called **dialysis**. During dialysis, the particles in a solution are separated on the basis of their size by using the difference in their rate of diffusion across a **selectively permeable membrane**. This type of membrane is one that permits the passage of certain substances and prevents the passage of others. For example, one type of selectively permeable membrane allows small molecules to pass through it but blocks the passage of large molecules.

Dialysis is a common procedure used to clear blood of urea and other toxic substances in cases of kidney failure. In this process, blood passes through a tube made of a membrane that is immersed in a solution similar to that of blood. In this exploration, an artificial dialysis membrane will be used. In living organisms, the cell membrane acts as a complex dialysis membrane.

**Facilitated diffusion** occurs when a molecule is transported across a membrane with the help of a carrier protein or a channel embedded in the membrane.

## Osmosis

**Osmosis** is a special type of diffusion that describes the spontaneous movement of solvent molecules across a selectively permeable membrane. In osmosis, the selectively permeable membrane permits the passages of solvent (water) molecules across the membrane but blocks the passage of solute molecules.

Motion occurs from a region where the solvent molecules are more concentrated to a region where they are less concentrated (Figure 2). Osmosis stops when equilibrium is reached by having equal concentrations on both sides of the membrane. For example, pickles are made by placing cucumbers in a concentrated brine solution. Water, the solvent, moves out of the cucumber, where it is more concentrated, and into the salty brine solution, where it is less concentrated, leaving us with a shriveled, shrunken pickle. On the contrary, a dried prune will swell when it is placed in water. The inside of the prune is more concentrated than the surrounding water, causing the surrounding water to flow into the prune by osmosis.

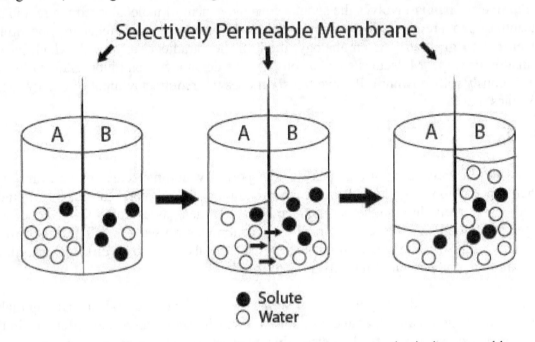

**Figure 2** Osmosis is the spontaneous flow of water across a selectively permeable membrane. In this figure, the solution is side A is more dilute (less solute, more water), and the solution in side B is more concentrated (more solute, less water). The water is diffusing from an area of high water concentration (A) to an area of low water concentration (B). Equilibrium is reached when there is no net movement of water from A to B.

If a red blood cell is placed in a solution of 0.15 M NaCl and there is no change in volume, this solution has the same number of nonpenetrating particles as the interior of the cell. In this case, the solution is **isotonic** to the cell. If the cell is placed in a 0.3 M NaCl solution, the cell will shrink (**crenate**) as water flows out of the cell into the solution, where it is less concentrated; this solution is **hypertonic** to the cell. In a **hypotonic** solution of 0.015 M NaCl, water enters the cell, causing it to swell and possibly burst (**hemolysis**). See Figure 3 for an illustration of these effects.

Tonicity is the same for all cell types in an organism, but it differs between species. For example, the tonicity of frog blood is different from that of human blood. To maintain the normal volume of isolated cells, or to replace blood by infusions of saline or other fluids, all such fluids must be isotonic to the cells. Physiological saline is 0.15 M NaCl. Clinically, a 0.9% NaCl solution is usually considered isotonic to human tissues. A 5% dextrose solution is also isotonic to human tissue and therefore is commonly used for intravenous therapy in hospitals.

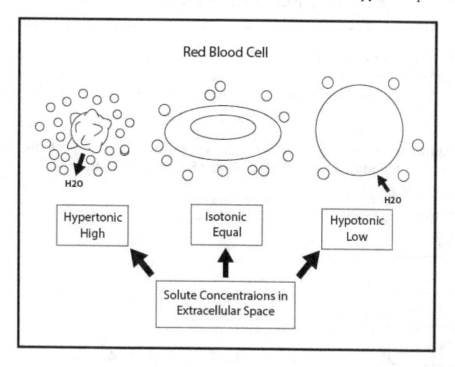

**Figure 3** The effect of solute concentration on the diffusion of water (osmosis) into and out of cells. In hypertonic conditions, substantial amounts of water diffuse out of the cell, causing crenation. In isotonic conditions, the diffusion of water out of the cell is equal to the diffusion of water into the cell. In hypotonic conditions, water diffuses into the cell; when too much water has entered the cell, it may cause lysis.

## Active Transport

**Active transport** involves the movement of molecules against a concentration gradient, which is an energy-requiring process. Active transport can be further classified as either **primary active transport** or **secondary active transport**. The energy used in primary active transport is provided by adenosine triphosphate (ATP), whereas the energy used in secondary active transport is provided by concentration gradients generated by primary active transport.

# EXPERIMENTAL PROCEDURES

You will do this exercise in two parts. In Part A, you will test the concept of osmosis by observing the physical properties of potato cylinders in different concentrations of salt solutions. In Part B, you will observe the diffusion of different-sized molecules through a semipermeable membrane.

**At the beginning of the lab session, fill the ceramic bowl on top of the hot plate with water. Turn on the hot plate—you need hot water for this experiment.**

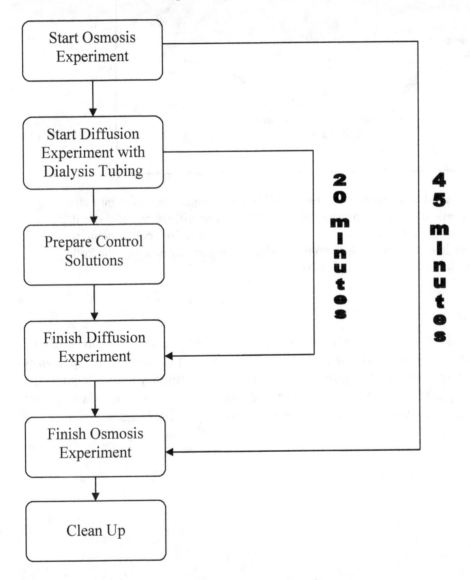

---

### SAFETY INFORMATION

__CAUTION__: **You must wear gloves and goggles for this experiment. Use both the Benedict's solution and I$_2$/KI solution with extreme care. These reagents are corrosive and irritating. Notify your lab instructor if you spill any on yourself.**

---

## PART A: OSMOSIS IN LIVING CELLS

1.  First, turn on your hot water bath, and allow it to come to a boil. This will be used in Part B of this lab exercise. You may need to refill it with water as it boils off. **DO NOT LET THE HOT WATER BATH BOIL DRY!**

2.  Use the cork borer at your lab bench to bore three cylinders from a raw potato. Remove the center metal rod of the borer first. Holding the potato horizontally, push the cork borer into the potato. You may want to turn the cork borer slightly to the left and to the right as you push it in; this may make it easier to release the potato cylinder. Push the potato cylinder out of the cork borer with the long metal rod.

3.  Using the scalpel provided on your lab bench, cut the three potato cylinders to approximately 40 mm in length. When you are done with the scalpel, place the sharp end into a piece of cork. Use the ruler provided to measure the length of each cylinder. Record the lengths (to the nearest 0.5 mm) in **Data Table 1** on your Data Sheets

4.  Use a pencil to label three weighing boats as **Cylinder 1, Cylinder 2, and Cylinder 3**. You must tare the empty weighing boat before placing the potato cylinder in it. Now, place each potato cylinder in its own weighing boat, and measure the mass of each potato cylinder to the nearest 0.01 g. Record the masses in **Data Table 1**.

5.  Observe the initial texture of each potato cylinder. Is it firm, soft, hard, pliable, flexible? Record your observations in **Data Table 1**.

6.  Pour approximately 25 mL of each solution listed below into the appropriately labeled 100 mL beaker. Each should be labeled with the name of the following three solutions:

    *   30% NaCl
    *   0.9% NaCl
    *   Distilled water

    Use the graduations on the beaker to estimate 25 mL.

7.  Place Potato Cylinder 1 into the 30% NaCl Solution, Potato Cylinder 2 into the 0.9% NaCl solution, and Potato Cylinder 3 into the distilled water. **Make sure there is enough of each solution in the beaker to COMPLETELY IMMERSE the potato cylinders.**

8. Set the timer on your lab bench for **45 minutes**. While the potato cylinders are soaking, you will proceed to Part B. You will return to Part A later in this exercise.

# PART B: DIALYSIS

> **Think of the dialysis tubing as the selectively permeable membrane of a cell, dividing the intracellular space from the extracellular space.**

1. Fill the large test tube about half full with distilled water, and place it in the tall beaker. Set it to the side for now.

2. Remove a soaked piece of dialysis tubing from the large glass beaker in the middle of your lab bench. Tie one end with a piece of string. To achieve a tight seal, make an approximately 1-cm fold at the end, and tie the string around the fold, knotting it firmly. Do not pull the string too tight or you will tear the tubing.

3. Use your fingers to separate the dialysis tubing from itself. You will have to rub the sides of the tubing up and down using your thumb and forefinger. Once you obtain an opening, use the large transfer pipette to open the tubing carefully along its entire length.

4. Shake your bottle of starch solution because much of the solute has settled to the bottom. Use the large transfer pipette to fill the dialysis tube halfway with the 0.25% starch solution.

5. Using a clean transfer pipette, fill the rest of the dialysis tube to within 2 inches of the top with 10% glucose solution. Make sure you leave at least 2 inches of space. Clamp the top of the dialysis tube with an orange closure, again folding it over slightly to achieve a tight seal.

6. Use the distilled water bottle held over a waste beaker to rinse any spilled glucose or starch solution from the outside of the tubing.

7. Place the filled dialysis tubing, tied end first, into the large test tube containing the distilled water (from Step 1). Hang the dialysis clamp from the large binder clip on the ring stand to keep the tubing upright and stable.

8. Set another timer for **20 minutes. During this time, you will prepare control solutions for the Benedict's (glucose) and iodine (starch) tests.**

9. Place the used pipettes in the large waste beaker at your lab bench.

You will use Benedict's solution to test for sugar and $I_2$/KI solution to test for starch. Figure 4 shows the differences between sugar and starch. Starch is a polymeric substance of varying molecular weight depending on the length of the chain, which can be either branched or unbranched. Glucose is the monomer from which we can build larger sugars and ultimately starch; it can be seen complexed with $I_2$/KI in Figure 5.

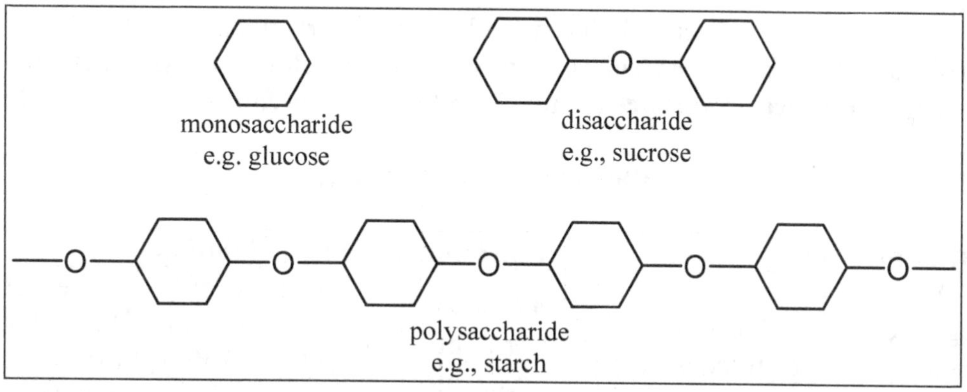

**Figure 4** Comparison of monosaccharides, disaccharides, and polysaccharides.

The Benedict's test allows us to detect the presence of **monosacharides,** all of which are reducing sugars. **Starch will not give a positive test with Benedict's solution** since it is a polymer, not a reducing sugar. The Benedict's test solution is blue due to the presence of copper sulfate ($CuSO_4$). In water, $CuSO_4$ will ionize to copper and sulfate ions; it is the copper ion that reacts with the sugar, producing a red-brown precipitate:

$$CuSO_4 \longrightarrow Cu^{2+} + SO_4^{2-}$$

$$2Cu^{2+} + \text{reducing sugar} \longrightarrow Cu^+$$

$$Cu^+ \longrightarrow Cu_2O \text{ (precipitate)}$$

The iodine test has long been used as a specific test for the presence of starch. The colorless solution will turn dark blue because starch in water adopts a helical configuration, and a row of iodine atoms fits neatly into the core of the helix as seen in Figure 5.

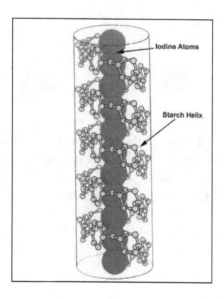

**Figure 5** $I_2/KI$ solution is a chemical indicator for the presence of starch because iodine atoms form a complex with starch by fitting into its hollow core.

10. Label 4 test tubes as in Table 1. Then, follow the instructions in Table 1 to make your **Control** solutions. **Your control solutions serve as references and will allow you to identify the presence or absence of starch and glucose in your experimental results.**

### Table 1: Control Solutions

| Test Tube | Preparation of Control Solution | | | |
|---|---|---|---|---|
| **1**<br>Negative Benedict's Control | 8 drops of distilled water | 4 drops of Benedict's reagent | Mix. Record the initial color in **Data Table 3**. | Place test tubes 1 and 2 in a test tube rack, and heat for 3 minutes. Remove and let cool. Record in **Data Table 3** the final color of each solution and whether glucose is present or absent. |
| **2**<br>Positive Benedict's Control | 8 drops of glucose solution | 4 drops of Benedict's reagent | Mix.<br><br>Record the initial color in **Data Table 3**. | |
| **3**<br>Negative Iodine Control | 8 drops of distilled water | 4 drops of I$_2$/KI solution | Mix.<br><br>Record the initial color in **Data Table 4**. | Record in **Data Table 4** the final color of each solution and whether starch is present or absent. |
| **4**<br>Positive Iodine Control | 8 drops of starch solution | 4 drops of I$_2$/KI solution | Mix.<br><br>Record the initial color in **Data Table 4**. | |

11. When 20 minutes have elapsed, test the **dialysate**, the solution surrounding the dialysis tubing, as well as the solution inside the dialysis tubing for the presence of glucose and starch. Label 4 test tubes as in Table 2. Follow Table 2 to test your solutions with the Benedict's and I$_2$/KI reagents. Record your observations in **Data Table 3**.

12. Answer Questions 3 and 4 on the Data Sheets.

**Table 2: Test Solutions**

| Test Tube | Testing Solutions Inside and Outside the Dialysis Tubing | | | |
|---|---|---|---|---|
| **5** Dialysate (Outside) | 8 drops of dialysate (Outside) | 4 drops of Benedict's reagent | Mix. Record the initial color in **Data Table 3**. | Place test tubes 5 and 6 in the test tube rack, and heat for 3 minutes. |
| **6** Solution (Inside) | 8 drops solution (Inside) | 4 drops of Benedict's reagent | Mix. Record the initial color in **Data Table 3**. | Remove and let cool. Record in **Data Table 4** the final color of each solution and whether glucose is present or absent. |
| **7** Dialysate (Outside) | 8 drops of dialysate (Outside) | 4 drops of I$_2$/KI solution | Mix. Record the initial color in **Data Table 4**. | Record in **Data Table 4** the final color of each solution and whether starch is present or absent. |
| **8** Solution (Inside) | 8 drops of solution (Inside) | 4 drops of I$_2$/KI solution | Mix. Record the initial color in **Data Table 4**. | |

# PART A: OSMOSIS IN LIVING CELLS (CONTINUED)

1. When 45 minutes have elapsed, use the forceps to remove the potato cylinders from their beakers.

2. **Thoroughly blot each potato cylinder with a paper towel—you must make sure that you remove ALL traces of liquid, or the mass of the potato cylinder will be flawed.**

3. Reweigh each cylinder to the nearest 0.01 g. Remember to zero the scale with the empty weighing boat before you weigh the cylinder. Record your results in **Data Table 2** on your Data Sheets.

4. Measure the final length of each of the three cylinders to the nearest millimeter. Record your results in **Data Table 2**.

5. Observe each cylinder's final texture, and record the results in **Data Table 2**.

6. Answer **Questions 1 and 2** on the Data Sheets.

# CLEAN-UP CHECKLIST

It is essential that you clean up properly after the experiments are complete.

❑ Turn off your hot water bath.

❑ Remember to cap all bottles.

❑ Discard the potato cylinders, used pipettes, used gloves, and used dialysis tubing into the trash.

❑ **Pour all solutions into correctly labeled waste containers at the back of the room. Any solutions containing I$_2$/KI must be rinsed into the I$_2$/KI waste container, and any solutions containing Benedict's solution must be rinsed into the Benedict's waste container.**

❑ **Discard the rinsed test tubes into the broken glass container.**

❑ Wipe up any spilled solutions with paper towels.

---

# DATA SHEETS:
## Diffusion and Osmosis

---

Name: _____ Date: _____

Lab Instructor: _____ Section: _____

## PART A: OSMOSIS IN LIVING CELLS

### Data Table 1: Osmosis Results BEFORE Soaking

| Solution | Potato Cylinders BEFORE Soaking | | |
|---|---|---|---|
| | Mass (g) | Length (mm) | Texture |
| **Cylinder 1** 20% NaCl | | | |
| **Cylinder 2** 0.9% NaCl | | | |
| **Cylinder 3** Distilled Water | | | |

### Data Table 2: Osmosis Results AFTER Soaking

| Solution | Potato Cylinders AFTER Soaking | | |
|---|---|---|---|
| | Mass (g) | Length (mm) | Texture |
| **Cylinder 1** 20% NaCl | | | |
| **Cylinder 2** 0.9% NaCl | | | |
| **Cylinder 3** Distilled Water | | | |

**Question 1:** The three NaCl solutions used in the osmosis portion of the experiment were of different tonicities relative to the solution inside the potato cells. Identify which solution was:

(a) Isotonic

(b) Hypotonic

(c) Hypertonic

**Question 2:** What type of solution (isotonic, hypotonic, or hypertonic) would you use for an intravenous injection of an antibiotic? Explain.

# PART B: DIALYSIS

## Data Table 3: Benedict's Tests

| Benedict's Test | Solution Color BEFORE Heating | Solution Color AFTER Heating | Glucose Present or Absent? |
|---|---|---|---|
| **Tube 1** (Negative Control) | | | |
| **Tube 2** (Positive Control) | | | |
| **Tube 5** (With Dialysate) | | | |
| **Tube 6** (With Solution Inside Tubing) | | | |

## Data Table 4: I$_2$/KI Tests

| I$_2$/KI Tests | Solution Color | Starch Present or Absent? |
|---|---|---|
| **Tube 3** (Negative Control) | | |
| **Tube 4** (Positive Control) | | |
| **Tube 7** (With Dialysate) | | |
| **Tube 8** (With Solution Inside Tubing) | | |

**Question 3:** Did large or small molecules diffuse through the dialysis tubing?

**Question 4:** What size molecules remained inside the dialysis tubing?

---

# EXPLORATION 17
# How Much Acetic Acid Is in Vinegar?

---

## Overview

In this lab, you will determine the acetic acid percentage in different commercial vinegars. The experiment involves use of an acid-base titration and careful analytic measurements.

## Preparation

You should be familiar with the following concepts and techniques:

- Components of vinegar
- Neutralization reaction
- Technique of titration
- End point of a neutralization reaction
- Role of an indicator
- Worked Example for an acid-base neutralization calculation

## Equipment and Supplies

Bring a calculator to this lab session.

**Equipment**
Burette
Ring stand
Ring stand clamp
Erlenmeyer flasks/beakers

**Supplies**
Stock NaOH solution
Phenolphthalein
Various commercial vinegars

# INTRODUCTION

## What Is Vinegar?

Commercial **vinegar** consists primarily of water and **acetic acid** ($CH_3COOH$). The chemical structure of acetic acid is shown in Figure 1. It is the acetic acid that gives vinegar its sharp taste and aroma. The word "vinegar" is derived from the French words for wine, *vin*, and sour, *aigre*. Thus, vinegar literally means "sour wine."

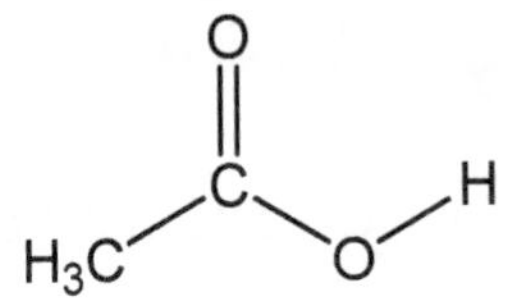

**Figure 1** Chemical structure of acetic acid ($CH_3COOH$).

Traditionally, vinegar is produced by a fermentation process that converts alcohol to form acetic acid and vinegar. This process can produce vinegar that contains concentrations of acetic acid greater than 20%. At this level, the vinegar is potentially corrosive and is unsafe for consumption. Commercial vinegars are diluted to a concentration of approximately 5% acetic acid by weight. In this exploration, you will measure the total acid concentration in a variety of different vinegars to test the authenticity of the acidity on the vinegar's label.

When an acid reacts with a base, the products are water and a salt. If a solution is acidic, it can be **neutralized** by adding a sufficient amount of base. In this exploration, you will also measure the acetic acid content of vinegar by performing a neutralization reaction with sodium hydroxide (NaOH). This reaction is shown below.

$$CH_3COOH(aq) + NaOH(aq) \rightarrow NaCH_3COO(aq) + H_2O(l)$$

The initial concentration of acetic acid can be determined by measuring the amount of sodium hydroxide required to neutralize the solution.

## Titration

The most common way to complete a neutralization reaction is through **titration**. A titration mixes two solutions, one of each reactant of a known chemical reaction, until the **end point** of the reaction is detected. The end point of the titration is the point at which both reactants (in this case, acetic acid and sodium hydroxide) are completely consumed. In an acid-base titration, the end point can be detected with an acid-base **indicator**, which is a compound that changes colors depending on the pH of the solution. In this experiment, you will be using the indicator phenolphthalein, which is colorless in acidic solutions and turns pink in basic solutions.

To carry out this titration, you will be using a burette to carefully measure the amount of NaOH added (or titrated) into a known amount of vinegar. The proper setup for the burette is shown in Figure 2.

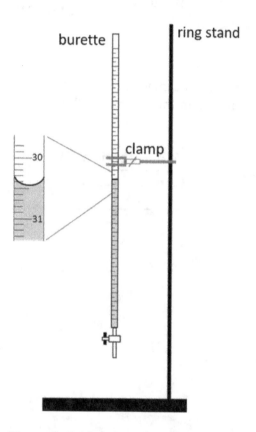

**Figure 2** Setting up and reading a burette.

The titration experiment depends on knowing how many moles of NaOH were necessary to completely neutralize the acetic acid in the vinegar sample. Review the Worked Example to learn how the amount of acetic acid is calculated from the titration results.

---

**WORKED EXAMPLE**

To determine percent acetic acid in vinegar from the titration, you first must determine how many moles of NaOH were added in the neutralization reaction. For example, if the concentration of NaOH is 0.250 mol/L and 19.98 mL were required to complete the reaction, then $4.995 \times 10^{-3}$ mol of NaOH were added to the vinegar.

$$\text{mol NaOH} = 19.98 \text{ mL} \times \frac{1 \text{ L}}{1000 \text{ mL}} \times \frac{0.2500 \text{ mol}}{1 \text{ L}} = 4.995 \times 10^{-3} \text{ mol}$$

Using the number of moles of NaOH and the molar ratio of NaOH to $CH_3COOH$ in the reaction (1:1), the mass of acetic acid (in g) in the vinegar sample can be determined.

---

$$\text{mass } CH_3COOH = 4.995 \times 10^{-3} \text{ mol NaOH} \times \frac{1 \text{ mol } CH_3COOH}{1 \text{ mol NaOH}} \times \frac{60.05 \text{ g } CH_3COOH}{1 \text{ mol } CH_3COOH}$$

$$= 0.2999 \text{ g } CH_3COOH$$

The percent acetic acid in vinegar is a mass ratio and thus is determined by the mass of acetic acid and the mass of vinegar in the sample. If the mass of vinegar was 6.00 g, then the % acetic acid is

$$\% \text{ acetic acid} = \frac{0.2999 \text{ g acetic acid}}{6.00 \text{ g vinegar}} \times 100 = 5.00\%$$

---

# EXPERIMENTAL PROCEDURES

---

### SAFETY INFORMATION

<u>CAUTION</u>: **Goggles and gloves must be worn for the duration of the experiment.**

---

## Preparing the Burette

Follow the procedure outlined below to properly clean and prepare your burette.

1. Rinse the burette 2 to 3 times with distilled water. Fill the burette approximately 1/5 full. Carefully tilt and rotate the burette so that the water coats the walls of the burette. Set the burette upright, and open the stop cock to allow the water to drain.

2. Repeat the above process with the NaOH solution. This will insure that any liquid remaining in the burette will be the titrant, not water. Thus, your titrant solution will not be diluted.

## Add the NaOH solution

1. Fill a burette with 50 mL of the standard NaOH solution.

2. Record the concentration of the stock NaOH solution in space provided in the Data Sheets.

## Prepare the Vinegar Solutions

1. Weigh a small beaker or Erlenmeyer flask.

2. Add approximately 5 mL of vinegar to the beaker, and reweigh the container.

3. Record these weights in **Data Table 1** on the Data Sheets.

4. Add one full dropper (~0.5 mL) of the indicator phenolphthalein to the vinegar solution. Swirl carefully to mix.

5. Label the beaker or flask with the name of the vinegar and the reported acetic acid concentration.

6. Repeat Steps 1 through 5 for each provided vinegar.

## Titration of Vinegar Solutions

1. Select one of your vinegar solutions, and place it under the burette.

2. Record the initial reading on the burette in **Data Table 2** on the Data Sheets.

3. Open the burette, and allow the NaOH to flow, dropwise, into the vinegar solution. Slowly stir the solution to mix.

4. Continue to add the NaOH until the solution turns a light pink color, and then close the burette. The pink solution should not fade in color after swirling the solution.

5. Record the final burette reading in **Data Table 2**.

6. Repeat the above procedure for the remaining vinegar solutions.

7. Answer **Questions 1 through 4** on the Data Sheets.

# CLEAN-UP CHECKLIST

It is essential that you clean up properly after the experiments are complete.

❑ All solutions used can be poured down the sink.

❑ Rinse all beakers and the burette with deionized water. The burette should be placed in the holder upside down to allow it to dry.

❑ Wipe down the ring stand and your work area, and discard paper towels in the trash.

---

# DATA SHEETS:
## How Much Acetic Acid is in Vinegar?

---

Name: _____ Date: _____

Lab Instructor: _____ Section: _____

## Concentration of Acetic Acid in Vinegar

Concentration of NaOH (mol/L): _____

### Data Table 1: Mass and Acetic Acid Concentration of Vinegar

| Vinegar Sample | Mass of Flask | Mass of Flask with Vinegar | Mass of Vinegar |
|---|---|---|---|
|  |  |  |  |
|  |  |  |  |
|  |  |  |  |
|  |  |  |  |
|  |  |  |  |

### Data Table 2: NaOH Titration of Vinegar

| Vinegar Sample | Initial Burette Reading | Final Burette Reading | Total Volume NaOH added | Final Concentration NaOH |
|---|---|---|---|---|
|  |  |  |  |  |
|  |  |  |  |  |
|  |  |  |  |  |
|  |  |  |  |  |
|  |  |  |  |  |

**Question 1:** Calculate the volume of NaOH added to each vinegar to neutralize the acetic acids.

**Question 2:** Determine the percent mass of the acetic acid in each vinegar.

**Question 3:** Determine the percent acidity of each vinegar.

**Question 4:** (a) You will likely find a difference between your calculated value and the percent acidity reported on the vinegar label. Use the equation below to calculate the **percent difference** between your calculated value and the label value:

$$\% \text{ difference} = \frac{(\text{label value} - \text{calculated value})}{\text{label value}} \times 100$$

(b) Suggest two reasons why your results differ from the acetic acid concentration reported on the label.

# EXPLORATION 18
## Extracting DNA from Cells

## Overview

The DNA in eukaryotic cells is enclosed within the nuclear membrane and surrounded by the cell's cytoplasm. To study or manipulate the DNA, it must first be extracted from the cell and separated from other cellular components. This lab experiment examines how DNA isolation is performed, with an emphasis on understanding the reason for each step in the procedure.

## Preparation

You should be familiar with the following concepts and techniques:

- Composition of eukaryotic cells
- Packaging of DNA into the cell nucleus
- Procedure for DNA isolation and the reason for each step

## Equipment and Supplies

Bring a calculator to this lab session.

**Equipment**
Beaker, 50 mL
Graduated cylinder, 25 mL
Funnel
Stirring rod

**Supplies**
Disposable pipettes
Cheesecloth
Plastic bag
Dosage cup
Sodium dodecyl sulfate (SDS), 10%
Phosphate-buffered saline (PBS), 1%
Ethanol
Saline, 1%
Bananas
Strawberries

---

# INTRODUCTION

---

## DNA in Eukaryotic Cells

Cells are traditionally classified into two basic types according to whether they possess a well-defined nucleus that encloses the cell's DNA. **Eukaryotic cells** (loosely translated as "good kernel") have such a nucleus, whereas **prokaryotic cells**, such as those of bacteria, do not. Since complex organisms, including humans, are constructed from eukaryotic cells, we need to examine how to extract the DNA from the cell nucleus.

The nucleus of a eukaryotic cell is enclosed by a double membrane called the **nuclear envelope**, which separates the nuclear components from the cytoplasm. Embedded within the cytoplasm are many other specialized internal structures called **organelles**, which include mitochondria, the endoplasmic reticulum, the Golgi apparatus, and lysosomes. A schematic diagram of a typical animal cell is given in Figure 1.

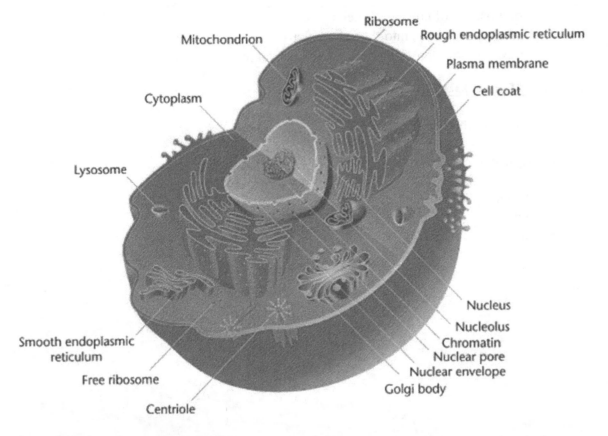

**Figure 1** A generalized animal cell showing several major organelles. The nucleus is the largest organelle and contains the DNA.

If the DNA strands within a single human cell could be uncoiled and laid end to end, they would extend to a length of 2 meters (over 6 feet). How can this length of DNA be packaged into the microscopic size of the cell nucleus? The answer is that the DNA is wound onto spools made of specialized proteins called **histones**. In turn, the histones are further organized into more complex structures to form a human **chromosome**. This efficient packing enables the space occupied by DNA to be shrunk by a factor of 50,000 compared to its extended length.

## Extracting DNA from Cells

In this exploration, you will perform a lab procedure to extract DNA from cells. First, you will extract the DNA from either bananas or strawberries. We use these fruits because they have a lot of DNA within each cell. As a follow-up experiment, you will employ the same procedure to obtain DNA from your cheek cells. Although you will obtain a smaller amount, it will perhaps be the first time that you will have seen your own DNA.

Figure 2 provides an overview of the procedure for extracting DNA from strawberries or bananas. Review this procedure and the reason for each step.

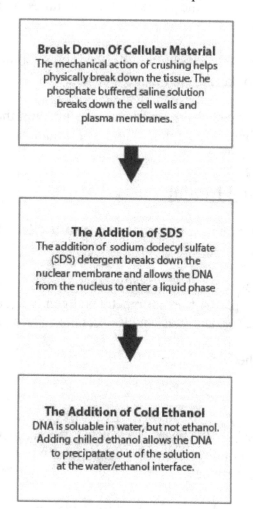

**Figure 2** Lab procedure for extracting DNA from bananas or strawberries.

# EXPERIMENTAL PROCEDURES

## PART A: ISOLATING DNA FROM BANANAS OR STRAWBERRIES

### Breaking Apart the Fruit Tissue

1. Place 50 g of assigned fruit (either banana or strawberry) into a plastic bag.

2. Seal the bag tightly, and crush the contents for 2 minutes or until the contents become a liquid slurry.

3. Measure 20 mL of cold phosphate-buffered saline (PBS) solution in a graduated cylinder. Add the buffer solution to the bag, and mix the contents until there is a uniform consistency.

4. Place a few layers of cheesecloth into a funnel, and position the funnel over a 50 mL glass beaker. After the blending is complete, filter the liquid slurry through the cheesecloth. Collect the filtrate in the beaker.

### Breaking Apart the Cell Nuclei

1. Take two 50 mL glass beakers, and label them **A and B**. Place 10 mL of the cell suspension into each beaker.

2. Add 5 mL of sodium dodecyl sulfate (SDS) solution to beaker **A**. Mix by drawing up the contents of the vial into a plastic transfer pipette and gently squeezing it out a few times. Do this carefully to avoid the formation of bubbles. Let the solution stand for a few seconds.

3. Answer **Question 1** on the Data Sheets.

### Extracting the DNA

1. Add 5 mL of SDS to beaker **B**. Mix by drawing up the contents of the vial into the pipette and gently squeezing it out a few times. Do this carefully to avoid the formation of bubbles. Let it sit on the bench for 1 minute.

2. After 1 minute, **CAREFULLY** add 15 mL (this volume can be approximate) of cold ethanol to each beaker. **This is best achieved by tilting the beaker and adding the ethanol along**

**the edge using a plastic transfer pipette**. If you add the ethanol carefully, you should notice two separate layers and an interface between them.

3.  Answer **Question 2** on the Data Sheets.

4.  Take a glass stirring rod, and dip one end into beaker **A**, through the layer of cold ethanol, into the SDS-nuclear suspension layer. Rotate the rod, and then raise it back into the layer of ethanol. Repeat this process until you have spooled the fibers onto the glass rod.

5.  Remove the rod, and examine the fibers wound around it. The fibers are the ethanol-insoluble form of DNA, bound together with some histone proteins.

6.  Answer **Question 3** on the Data Sheets.

---

**After you have completed you DNA extraction and analysis, compare your results with your partner's results generated by using the other fruit tissue.**

---

## PART B: ISOLATING DNA FROM CHEEK CELLS

### Collecting Cheek Cells

1.  Rinse out your mouth in the sink with water using a dosage cup. Please make sure that you have removed all food and beverage residue from the inside of your mouth.

2.  Once you have cleaned out your mouth, fill the same cup with a 1% saline solution, and rinse your mouth vigorously for 30 seconds. Spit the solution back in the cup.

3.  Transfer 10 mL of the saline rinse to a 50 mL beaker.

### Breaking Apart the Cell Nuclei

1.  Add 5 mL of the SDS solution to the beaker. Mix by drawing up the contents of the beaker into the pipette and gently squeezing it out a few times. Do this carefully to avoid the formation of bubbles. Let the solution stand for 1 minute.

### Extracting the DNA

1.  After 1 minute, **CAREFULLY** add 15 mL (this volume can be approximate) of cold ethanol to each beaker. **This is best achieved by tilting the beaker and adding the ethanol along the edge using a plastic pipette.**

2.  Take a glass stirring rod, and dip one end into the beaker. The rod should pass through the layer of cold ethanol, into the SDS-nuclear suspension layer. Rotate the rod, and then raise it

back into the layer of ethanol. Repeat this process until you have spooled the fibers onto the glass rod.

3.  Remove the rod, and examine the fibers wound around it. The fibers are the ethanol-insoluble form of DNA, bound together with some histone proteins.

---

# CLEAN-UP CHECKLIST

---

It is essential that you clean up properly after the experiments are complete.

❑ Dispose all DNA samples in the waste container in the back of the lab.

❑ Clean the glassware, and set on a paper towel to dry at your lab bench.

❑ Discard of pipettes in the trash can.

❑ Wipe up any spills on the bench.

❑ Remove your gloves, and dispose of them in the trash can. Dispose of any other miscellaneous waste in the trash can.

---

# DATA SHEETS:
## Extracting DNA from Cells

---

Name: _____ Date: _____

Lab Instructor: _____ Section: _____

## PART A: ISOLATING DNA FROM BANANAS OR STRAWBERRIES

**Question 1:** Compare the appearance and consistency of the solution **with** the SDS to the one **without** it. How did the appearance, viscosity, and consistency of the nuclear suspension change after the addition of the SDS? Explain what happens upon addition of SDS solution.

**Question 2:** What do you see at the interface? Explain your observations based on what you learned in the introduction about the effect of ethanol on DNA.

**Question 3:** Describe the appearance of these fibers.

# EXPLORATION 19
# Separating Proteins by Column Chromatography

## Overview

Our cells contain many thousands of different biological molecules. To investigate a specific molecule such as an individual protein, we must find some way to separate it from all of the others. A wide range of **separation techniques** can be used to achieve this goal. In this exploration, you will use one of these methods, called **column chromatography**, to separate molecules of differing size and estimate the size of a protein.

## Preparation

You should be familiar with the following concepts and techniques:

- Size exclusion chromatography
- Molecular weight of proteins
- Mobile and stationary phases of a column
- Role of gel particles in the separation process
- Sample elution and collection of fractions

## Equipment and Supplies

Bring a calculator to this lab session.

**Equipment**
Chromatography columns
Sephacryl chromatography beads
Plastic cuvettes
Cuvette holders

**Supplies**
Blue dextran dye
Cytochrome *c*
Catalase
Hydrogen peroxide
Sharpie pen

---

# INTRODUCTION

---

## What Is Chromatography?

Chromatography is a process used to separate molecules on the basis of a chemical property, such as size, charge, or solubility. The word is derived from the Greek *chroma* (meaning "color") because the earliest experiments in chromatography involved the separation of different color dyes on fine sand (silica). A mixture of dyes was dissolved in a solvent, the colored solution was applied to a column filled with silica, and then the solvent was slowly passed through the column. The different colors appeared as separate bands and could be collected in separate flasks.

Today, many different types of chromatography are used in labs to purify and separate a wide variety of molecules. This technique continues to be effective for isolating and purifying all types of biomolecules. In this exploration, you will use **size exclusion chromatography** to separate proteins based on their size. This type of separation process is also called **gel filtration chromatography**.

## Separating Proteins by Size

The size of a protein is usually given in terms of its **molecular weight**, which uses the hydrogen atom as the unit of reference. Biological molecules range in molecular weight from less than a hundred to as large as several million. Size exclusion chromatography, which is capable of separating larger biological molecules, has been used to purify thousands of different proteins, nucleic acids, enzymes, sugar polymers, and other biomolecules. In addition, the technique has been applied to the molecular weight determination of unknown substances by comparison to molecules of known molecular weight.

All types of chromatography are based on similar basic principles. The sample to be examined is allowed to interact with two physically distinct entities: a **mobile phase** and a **stationary phase**. The mobile phase is the solvent that moves through the column, which in our case is a phosphate buffer at pH 7. The stationary phase is the column packing, which has the ability to selectively interact with the molecules that are moving through the column.

Figure 1 shows an illustration of the apparatus you will use in this exploration. In size exclusion chromatography, the stationary phase consists of small **gel particles** that contain small **pores** of a controlled size. A solution containing the sample mixture of various molecular sizes passes through the column with continuous solvent flow. Smaller molecules can fit inside the pores and so they enter the gel particles. As a result, these molecules travel a longer and more convoluted path as they pass through the columns. By contrast, sample molecules larger than the pores cannot enter the interior of the gel beads, so they are excluded. Molecules that are excluded

216

from the beads will pass through the column at a faster rate because the distance they have to travel is shorter. The different paths of each molecule through the column are illustrated in Figure 2.

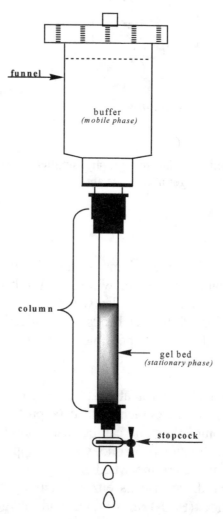

**Figure 1** The apparatus for size exclusion chromatography. The column is packed with gel particles that contain small pores. The stopcock opens a small valve that allows the buffer to flow through the column. Here, the valve is shown in the open position.

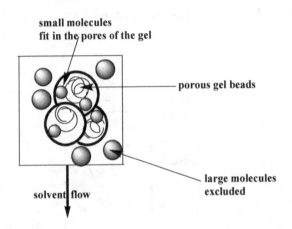

**Figure 2** During size exclusion chromatography, smaller molecules fit within the pores of the gel particles, whereas larger molecules are excluded.

When the molecule emerges from the column, we say that it has **eluted** (from the Latin word meaning "washed out"). The larger molecules will elute first because they have taken a shorter path through the column. The smaller molecules will elute later because they have followed a longer path through the interior of the gel beads. By collecting successive **fractions** of the mobile phase (e.g., every 0.5 or 1 mL), it is possible to capture the eluting molecules in different volumes, thereby achieving our goal of separation.

You will first use size exclusion chromatography to separate a mixture of two molecules: **blue dextran** (a sugar polymer) and **cytochrome** *c* (a protein). These molecules have been selected because they have strong colors and can be easily observed both while moving through the column and in the collected fractions. Second, you will apply a sample of **catalase**, a protein that functions as an enzyme, to determine whether it is larger or smaller than cytochrome *c*. Since catalase is weakly colored, we use its enzyme activity to detect its presence. Catalase decomposes hydrogen peroxide ($H_2O_2$) into water and oxygen ($O_2$). By adding drops of hydrogen peroxide to each of the collected fractions, the presence of catalase will be revealed by bubbles of oxygen gas.

---

# EXPERIMENTAL PROCEDURES

---

## PART A: SEPARATING MOLECULES BY COLUMN CHROMATOGRAPHY

### Preparing the Cuvettes

1. Label 12 plastic cuvettes as **1 through 12**, and place them in a rack. You will use these to collect your 0.5 mL fractions.

2. Pipette 0.5 mL of buffer into another cuvette. Draw a line at the 0.5 mL mark. You will use this cuvette as your 0.5 mL volume standard. Using a sharpie pen and your standard, draw a line indicating the 0.5 mL mark on each of your 12 labeled cuvettes. This will help you to measure the 0.5 mL fractions more accurately during the experiment.

### Preparing the Sample

1. Label 12 plastic cuvettes as **1 through 12**, and place them in a rack. You will use these to collect your 0.5 mL fractions.

2. Using a glass Pasteur pipette, add **2 drops** of blue dextran to a small tube.

3. Pipette **3 drops** of cytochrome *c* into the same tube.

4. Gently tap the tube so that all of the sample will be mixed together at the bottom.

5. Carefully pick up **4 to 5 drops** with a Pasteur pipette, and apply this to the column according to the next set of instructions.

### Loading the Sample on the Column

Figure 3 illustrates the process of loading the sample on the column. Before you proceed, **read the text for description in full** since the lettered paragraphs correspond to the steps shown in the figure. The sample should be applied in the **narrowest possible band** at the top of the column as shown.

> **During this experiment, make sure that the column is never allowed to dry out. Always keep the column bed bathed in the buffer.**

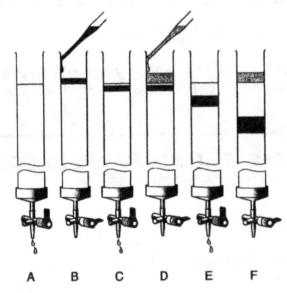

**Figure 3** The technique for loading samples onto the column.

A. **Place a waste beaker under the column**. Allow the buffer to flow until its level reaches the top of the settled bed, but without allowing the bed to dry. If this step is taking too long, you can speed up the process by removing the excess buffer with a long plastic pipette. When the level of the buffer is slightly above the top of the gel bed, close the valve to stop the flow. You can then remove the remaining buffer using a Pasteur pipette.

B. Add the sample gently to the gel bed. Touch the pipette tip to the glass surface at the top of the gel bed, and make a **circular motion** with the tip as you slowly eject the sample, layering it evenly on the top of the column bed. Try to avoid disturbing the settled bed.

C. After the sample has been loaded to the top of the column, open the valve at the bottom, and let the sample flow into the column. Close the valve once the last of the sample enters the gel bed.

D. Gently add a small volume of elution buffer (about equal to the volume of the sample) to the top of the column, using the same technique used for loading the sample.

E. Open the valve at the bottom, and allow the buffer to enter the column as you did with the sample, again closing the valve before the column runs dry. You will see the color move slightly down into the gel bed.

F. Finally, add a larger quantity of buffer over the gel bed. Do this **gently at first** so that you do not disturb the bed. When you have almost filled the column with buffer, connect the reservoir to the top. Gently add enough buffer to the reservoir until it is approximately two-thirds full. You are now ready to run your column.

## Collecting the Fractions

1. Remove the waste beaker, and place **cuvette 1** under the column. There should be only a small distance between the stopcock and the cuvette. If the stopcock is too high, carefully lower the column by moving the clamp.

2. Fully open the stopcock. After 0.5 mL has collected in the cuvette, quickly switch to **cuvette 2**, placing **cuvette 1** back in the rack. Do not close the stopcock between collecting fractions, and try not to lose drops while switching the cuvettes.

3. Continue collecting 0.5 mL fractions until you have 12 samples. Close the stopcock to prevent any further flow of liquid.

4. Before analyzing your results, **begin washing your column** with buffer in preparation for Part B of the lab project. This is accomplished by placing a waste beaker under the column, opening the stopcock, and allowing the buffer to run through for 10 minutes. **Make sure that you have sufficient buffer in the reservoir. While the column is being washed, move on to Step 5.**

5. Hold each of the 12 fractions against a white sheet of paper, and carefully look for any visible color to the solution. This is often best achieved by looking down at the sample from above. Enter your observations in **Data Table 1** and answer **Questions 1 to 3** on the Data Sheets.

# PART B: IS CATALASE LARGER OR SMALLER THAN CYTOCHROME *c*?

## Loading the Catalase Sample

1. Label a new set of cuvettes as **1 through 12**. For each cuvette, use your standard to draw a line indicating a volume of 0.5 mL.

2. Stop the column washing procedure by closing the stopcock. Remove the bulk of the remaining buffer in the reservoir using a large plastic pipette. When only a small volume of buffer is left, switch to using the glass Pasteur pipette to remove the remaining liquid.

3. Using a Pasteur pipette, carefully pick up and load **4 to 5 drops** of the catalase solution onto the column. **Follow the instructions in Part A for sample loading.** The final stage should be to attach the reservoir and fill it approximately two-thirds full with buffer.

4. Fully open the stopcock. After 0.5 mL has collected in **cuvette 1**, quickly switch to cuvette 2. Do not close the stopcock between collecting fractions. Try not to lose drops while switching the cuvettes.

5. Continue collecting 0.5 mL fractions until you have 12 samples. Close the stopcock to prevent any further flow of liquid.

6. **Wash your column again with buffer for 10 minutes.** Watch the column periodically to make sure that the column does not run dry. **While the column is being washed, more on to Step 7.**

7. Hold each of the 12 fractions against a white sheet of paper, and carefully look (from above) for any visible color to the solution. Add your observations to **Data Table 2** on the Data Sheets.

8. **Put on gloves and goggles for this next step.** Using a Pasteur pipette, add **3 drops** of $H_2O_2$ to each of the fractions. Note any bubbling due to the release of oxygen, and enter your observations in **Data Table 2**. Answer **Questions 4 and 5** on the Data Sheets.

---

# CLEAN-UP CHECKLIST

---

It is essential that you clean up properly after the experiments are complete.

☐ Wash out both the smaller practice column and the longer column with buffer. When the column has cleared, close the stopcock on the column to stop the flow of buffer. **MAKE SURE THERE IS STILL PLENTY OF BUFFER ON THE COLUMN TO KEEP IT FROM DRYING OUT.**

☐ Pour the liquid in the waste beaker down the sink (since it contains no harmful material).

☐ Wipe up any spills with a paper towel.

☐ Dispose of used cuvettes, small pipettes, gloves, and other related waste in the garbage. **Do not throw away the large plastic pipettes.**

☐ Place the used glass Pasteur pipettes in the glass box. **DO NOT DISPOSE OF ANY OTHER ITEMS IN THE GLASS DISPOSAL BOX.**

---

# DATA SHEETS:
## Separating Proteins by Column Chromatography

---

Name: _____ Date: _____

Lab Instructor: _____ Section: _____

## PART A: SEPARATING MOLECULES BY COLUMN CHROMATOGRAPHY

Add your observations for each fraction in Data Table 1. Include both the color (red or blue) and its intensity using the following system:

+ weak color          ++ moderate color          +++ strong color

### Data Table 1: Column Fractions for Blue Dextran and Cytochrome *c*

| Fraction | Color | Fraction | Color |
|----------|-------|----------|-------|
| 1 | | 7 | |
| 2 | | 8 | |
| 3 | | 9 | |
| 4 | | 10 | |
| 5 | | 11 | |
| 6 | | 12 | |

**Question 1:** Based on the results in Data Table 1, which fractions contained blue dextran, and which fractions contained cytochrome *c*?

Fractions for blue dextran: _____

Fractions for cytochrome *c*: _____

**Question 2:** Did you achieve good separation with the column (i.e., were blue dextran and cytochrome *c* divided into distinct fraction ranges)?

**Question 3:** Which molecule has the higher molecular weight, blue dextran or cytochrome $c$? Explain your answer based on the method of size exclusion chromatography.

# PART B: IS CATALASE LARGER OR SMALLER THAN CYTOCHROME *c*?

Add your observations for each fraction in Data Table 2.

- Use YES or NO to indicate if you can see any color in the sample.

- We measure the activity of catalase by observing the release of $O_2$ gas when hydrogen peroxide is added to the fraction. Characterize the activity using the following scale:

    − no activity     + weak activity     ++ moderate activity     +++ strong activity

**Data Table 2: Column Fractions for Catalase**

| Fraction | Color | Activity | Fraction | Color | Activity |
|----------|-------|----------|----------|-------|----------|
| 1 | | | 7 | | |
| 2 | | | 8 | | |
| 3 | | | 9 | | |
| 4 | | | 10 | | |
| 5 | | | 11 | | |
| 6 | | | 12 | | |

**Question 4:** Based on your observations, which fractions contain catalase?

   Fractions for catalase: _____

**Question 5:** Compare the fractions containing catalase to those containing cytochrome $c$ in the previous experiment. Which is the larger protein, catalase or cytochrome $c$? Explain the reason for your answer.

# EXPLORATION 20
# Measuring the Enzyme Activity of Catalase

## Overview

Many of the key biological functions in our bodies are performed by **enzymes**. An enzyme is a biological molecule—usually a protein—that acts as a **biological catalyst** by increasing the rate of a chemical reaction without being permanently changed in the process. This lab project investigates the activity of enzymes using **catalase**, which accelerates the conversion of hydrogen peroxide into water and oxygen gas.

## Preparation

You should be familiar with the following concepts and techniques:

- Enzymes as biological catalysts
- Substrates and the active site
- Biological function of catalase
- Turnover number
- Enzyme inhibitor

## Equipment and Supplies

Bring a calculator to this lab session.

**Equipment**
Paring knife
Scale
Blender
Funnel
Ring stand and clamps
Stir plate
Stir bar
Graduated cylinder, 25 mL
Stopper
Finnpipette, 1–5 mL
Plastic bin

**Supplies**
Potato
Cheesecloth
Weigh boat
Centrifuge tubes, 15 mL
Phosphate buffer, pH 7
Test tubes
Tubing
Hydrogen peroxide, 2%
Methanol

# INTRODUCTION

## Catalase: An Example of a Biological Catalyst

**Enzymes** serve critical functions in the body, catalyzing nearly every reaction that takes place. We can see the power of enzyme catalysis in the decomposition of **hydrogen peroxide** into water and oxygen:

$$2H_2O_2 \;\rightarrow\; 2H_2O + O_2$$

This reaction is normally very slow. You can purchase a bottle of $H_2O_2$ solution and keep it on the shelf for many months before it decomposes. If, however, you apply hydrogen peroxide solution to a cut finger, you will notice an immediate bubbling from released oxygen gas. This effect is produced by **catalase**, an enzyme carried in many cells that **increases the rate of peroxide decomposition about 1 billion fold.** The catalase enzyme has evolved to defend cells against the damaging effects of $H_2O_2$ since this molecule is a reactive and potentially harmful product of certain chemical reactions.

The overall structure of catalase is shown in Figure 1. It contains four protein subunits, and each protein subunit in turn contains a chemical group called a **heme**. In general, enzymes function by binding one or more reactant molecules called **substrates**. In the case of catalase, the substrate is the hydrogen peroxide ($H_2O_2$) molecule. The catalyzed reaction takes place in a region of the enzyme called the **active site**. For catalase, the heme groups function as the active sites.

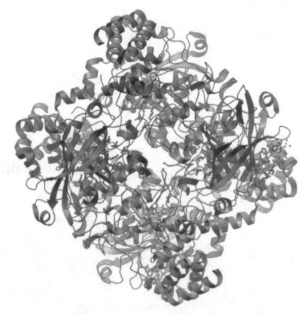

**Figure 1** The structure of catalase contains four subunits. Each subunit contains a heme group that functions as the enzyme's active site.

Within the active site, the $H_2O_2$ substrate is rapidly converted into product molecules, water and oxygen gas, which are then released by the enzyme. The active site contains amino acid sidechains that are arranged in a precise configuration to facilitate the chemical reaction. Some enzymes also utilize a small accessory molecule, called a **coenzyme** or **cofactor,** to assist with the chemical reaction. Catalase is one of these enzymes, and its cofactor, called a **heme** group, contains an iron atom in the center of a cyclic molecule. Hemes are found in many other enzymes and in the proteins that transport oxygen in our blood.

Catalase is one of the most efficient biological catalysts known. A true catalyst participates in the reaction yet is **unchanged** by it. After catalyzing the decomposition of one $H_2O_2$ molecule, catalase is found in exactly the same state as before, ready for another catalytic reaction. The efficiency of an enzyme is given by its **turnover number**, which is defined as the number of substrate molecules (reactant molecules) that one enzyme unit can transform in 1 minute. At body temperature, the turnover number for catalase is 5.6 million. This means that one catalase unit can bind and decompose 5.6 million $H_2O_2$ molecules each minute, or approximately 90,000 molecules per second.

## Enzyme Inhibitors

There is an important class of drugs called **enzyme inhibitors**. As their name suggests, they are molecules that inhibit the function of an enzyme by slowing its activity. Most enzyme inhibitors bind to the active site of the enzyme and prevent the substrate from binding. Many familiar drugs, such as aspirin, Lipitor, and Viagra, function as enzyme inhibitors.

During this exploration, you will investigate the effect of an enzyme inhibitor on the function of catalase. This experiment is a model for enzyme inhibitors that are used as pharmaceuticals.

---

# EXPERIMENTAL PROCEDURES

---

### SAFETY INFORMATION

<u>CAUTION</u>: Even at a concentration of **2%**, hydrogen peroxide is harmful to your eyes and skin. Goggles and gloves must be worn for the duration of the experiment.

---

## PART A: PREPARING THE POTATO EXTRACT

1. Retrieve a potato, cheesecloth, two 15 mL conical tubes, and weigh boat from the front bench.

2. Chop the potato into small pieces (about the size of a cube that you would cut cheese into for serving with crackers). **Be careful not to cut yourself.**

3. Place the plastic weighing tray on the electronic balance, and set the measurement to zero by taring it. Use the balance to measure approximately **50 grams** of chopped potato (this measurement does not need to be precise).

4. Place the 50 g of chopped potato in the blender. Measure 40 mL of pH 7 phosphate buffer using a graduated cylinder, and then pour it in the blender. This buffer stabilizes the solution pH. Blend for **30 seconds** or until the potato has reached the consistency of a piña colada.

5. While the potato is blending, place a large funnel into a clean beaker. Cover the funnel with a piece of clean cheesecloth that has been folded several times.

6. Pour the blended potato into the cheesecloth. Wrap the cheesecloth securely around the potato, and squeeze the potato juice through the funnel. Make sure that the cheesecloth is securely wrapped or else solids will contaminate the potato extract.

7. Carefully pour the potato juice from the beaker to fill two 15 mL plastic centrifuge tubes, taking care not to disturb the solid at the bottom of the beaker. **Place the tubes immediately on ice.** Each group will use a tube of this juice as their catalase source for the experiments.

## PART B: MEASURING CATALASE ACTIVITY

You will monitor the activity of the catalase enzyme by measuring the oxygen gas that is produced from the decomposition of hydrogen peroxide ($H_2O_2$). The apparatus shown in Figure 2 will be used to measure the volume of oxygen since the oxygen will displace water from the inverted graduated cylinder. You will determine how much oxygen is produced in a set amount of time.

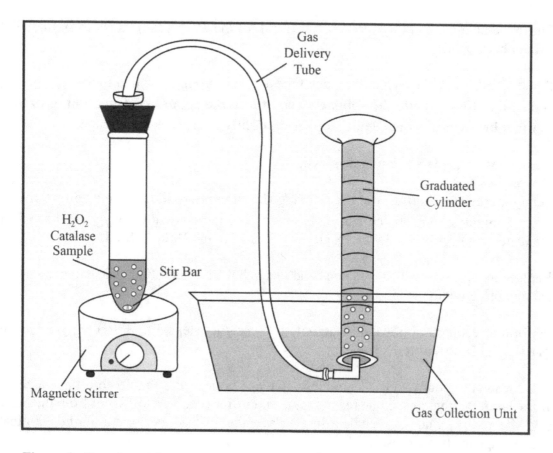

**Figure 2** Experimental apparatus to measure catalase activity by collecting the oxygen gas (seen here as circles entering the graduated cylinder) produced by the decomposition of hydrogen peroxide.

1. Fill the 25 mL graduated cylinder with tank water by completely submerging the cylinder in the tank. Keep the cylinder under water, and tip its lip slightly upward to allow any air bubbles to escape. Invert the cylinder, making sure that the lip stays under water, and secure it with a clamp. **You will need to submerge and refill the cylinder with tank water between each round of the experiment.**

2. Since the graduated cylinder is **inverted**, you will be reading the markings **upside down**.

3. Look at Figure 2 to understand how you will connect the **test tube** to the **inverted graduated cylinder** with rubber tubing. On one end of the rubber tubing is an L-shaped connector. Place this into the inverted graduated cylinder, and make sure it fits securely so that all of the gas will enter the graduated cylinder. The other end with the rubber stopper will be inserted into the test tube to the sidearm tube once the reaction occurs.

*NOTE:* **INFORM THE LAB INSTRUCTOR WHEN YOU REACH THIS POINT SO THEY CAN VERIFY YOUR SETUP IS CORRECT.**

4. Place a small, clean magnetic stir bar into a clean test tube.

5. Shake the potato extract well, and pipette 2 mL of the extract into the sidearm tube. Put the extract back on ice.

6. Pipette 2 mL of 2% $H_2O_2$ into the same tube and seal promptly with a rubber stopper. When adding the $H_2O_2$, **pipette the volume straight into the catalase extract, not onto the side of the tube**. Remember to put the **stopper** on **tightly** or else gas will escape.

7. Turn the stir plate **ON** to a middle speed.

8. Set a timer for **5 minutes. Start measuring the time immediately after adding the $H_2O_2$.** After 5 minutes, turn off the stir plate, and record the amount of oxygen collected in the graduated cylinder. Record your results as "Trial 1" in the **Data Table 1.**

9. Replace the test tube with a clean one and reattach the rubber tubing. Clean the magnetic stir bar and place it into the tube.

10. Perform a second trial of catalase activity by repeating steps 5 to 8. Record your results as "Trial 2" in **Data Table 1**.

11. If the results of your two trials are close, average your results, and record this value on the **Data Table 1**. If you have one result that is an **outlier** (i.e., very different from the other by, say, a factor of 2), do not include it in your average. Instead, perform a third determination, and average the two closest results.

12. Answer **Question 1** on the Data Sheets.

## PART C: EFFECT OF AN ENZMYE INHIBITOR

In this exercise, you will investigate how much of an enzyme inhibitor is required to depress enzyme activity. We will be using **methanol** as the inhibitor and will test volumes of 20, 50, 100, and 200 µL to study the reduction of enzyme activity over a range of inhibitor amounts. Because you are sharing a potato extract mixture, you will be able to share data with the other group at your table. One group should test volumes of 20 and 100 µL, and the other group should test 50 and 200 µL. Discuss as a table which volumes each group will be testing.

1. Gently shake up your catalase tube to evenly distribute the juice.

2. Using a **clean pipette tip**, add 2 mL of the catalase to a clean sidearm tube. Put the extract back on ice.

3. With a pipette, add your first volume (20 or 50 µL) of methanol. **Please note that this is a very small volume**. Make sure that you add the inhibitor directly into the catalase

preparation, not along the side of the tube. Gently swirl your sidearm tube to ensure the inhibitor is mixed into the solution.

4. Pipette 2 mL of 2% $H_2O_2$ into the same tube, and seal promptly with a rubber stopper. When adding the $H_2O_2$, **pipette the volume straight into the catalase extract, not onto the side of the tube**. Remember to put the **stopper** on **tightly** or else gas will escape.

5. Turn the stir plate **ON** to a middle speed.

6. Set a timer for **5 minutes. Start measuring the time immediately after adding the $H_2O_2$**. After 5 minutes, turn off the stir plate, and record the amount of oxygen collected in the graduated cylinder. Record your results on the **Data Table 2** on the Data Sheets.

7. Repeat the experiment using your second volume (100 or 200 µL) of methanol. Record all results in **Data Table 2**.

8. Answer **Question 2 and 3** on the Data Sheets.

---

# CLEAN-UP CHECKLIST

---

It is essential that you clean up properly after the experiments are complete.

❏ Using the stir bar retriever, retrieve the small stir bar from the test tubes, wash with water, and return to the instructor's bench.

❏ Pour all waste from the blender, test tubes, and conical tubes down the sink.

❏ Wash the blender, test tubes, stir bar retriever, funnel, beaker, cutting board, and knife with hot water, but do not use any soap (since this will affect the enzyme activity).

❏ Throw out all pipettes tips, extra potato, cheesecloth, conical tubes, and weigh boats in the trash can.

❏ Deassemble the setup.

❏ Turn **OFF** the stir plates.

## DATA SHEETS:
## Measuring the Enzyme Activity of Catalase

Name: _____ Date: _____

Lab Instructor: _____ Section: _____

## PART B: MEASURING CATALASE ACTIVITY

For each trial, **record the volume of $O_2$ produced in 5 minutes**.

### Data Table 1: Measurements of Catalase Activity

| Enzyme Activity | |
|---|---|
| Trial | Volume (mL) |
| 1 | |
| 2 | |
| 3 (If Needed) | |
| Average | |

**Question 1:** Did your first two measurements give you similar results? Or did you have an "outlier" in your first two trials (i.e., a difference in results by a factor of 2)? If so, specify which measurement was the outlier.

# PART C: EFFECT OF AN ENZMYE INHIBITOR

For each trial, **record the volume of $O_2$ produced in 5 minutes.** For the first data point **with no inhibitor added**, use your average result obtained from Part B.

Share your data with the other group at your table. This will allow you to analyze a full data set with only needing to test two volumes.

### Data Table 2: Measurements of Catalase Activity in the Presence of an Enzyme Inhibitor

| Enzyme Inhibitor | |
|---|---|
| **Inhibitor Volume** | **$O_2$ Volume (mL)** |
| 0 μL (No Inhibitor) | |
| 20 μL | |
| 50 μL | |
| 100 μL | |
| 200 μL | |

**Question 2:** Create a graph of your data using the graph paper below.

- Use the left-hand and bottom edges of the graph paper as your graph axes. Use the following labels for the axes:

    Horizontal axis = volume of inhibitor ($\mu L$)

    Vertical axis = volume of $O_2$ gas (mL)

- Create a title for your graph, and write it above the graph paper.

- Add **your data points** from Data Table 2.

- Join the data points with **straight lines**.

**GRAPH TITLE:**

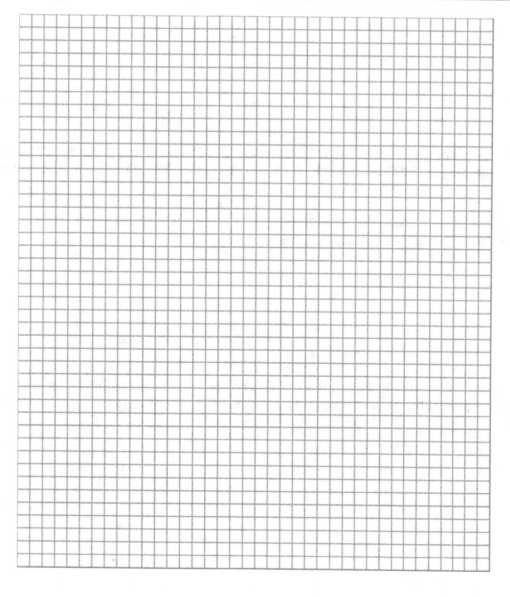

**Question 3:** The presence of an enzyme inhibitor reduces the activity of the enzyme. We can make this statement more precise by calculating the **percent inhibition** for a specific amount of inhibitor. As a measure of the activity, we will use the volume of oxygen collected after 5 minutes.

(a) For the enzyme inhibitor experiment, calculate the percent inhibition for both **20 and 200 μL** using the equation below. Recall that the use of |.....| indicates the **absolute value** (i.e., the number without the sign). Record your results in the spaces below.

$$\% \text{ inhibition} = \frac{(\text{volume of O}_2 \text{ without inhibitor} - \text{volume of O}_2 \text{ with inhibitor})}{\text{volume of O}_2 \text{ without inhibitor}} \times 100$$

  20 μL    % inhibition = _____

  200 μL   % inhibition = _____

*Note:* **If you were not able to complete the 200-μL inhibitor experiment for any reason, calculate the percent inhibition for the LARGEST AMOUNT of inhibitor that you used.**

(b) Compare your percent inhibition results at 20 and 200 μL. Explain the difference in terms of the action of the inhibitor on the enzyme.

(c) Based on your results in Parts (a) and (b), **predict** how much inhibitor you would need to produce **100% inhibition of the catalase enzyme** (i.e., completely stop its function).

# EXPLORATION 21A
# Bacteria and Antibiotics (Part 1)

## Overview

This lab project provides an opportunity for you to investigate **bacteria** and **antibiotics**. Bacteria are all around us—on our skin, on the surfaces we touch, and in our digestive system. But these bacteria are so small that we can't observe them directly, so we usually pay little attention to them until they affect our health. In this exploration, you will become familiar with some of the methods of studying microbes in the labor and investigate how antibiotics affect the growth of different types of bacteria.

## Preparation

You should be familiar with the following concepts and techniques:

- How are bacteria grown in a lab
- How Gram is staining used to identify different types of bacteria
- The feature of a cell that determines if it is Gram positive or negative
- The connection between Gram staining and the effect of antibiotics

## Equipment and Supplies

**Equipment**
Staining dish and rack
Hot plates
Forceps
Desk lamp
Microscopes
Incubator

**Supplies**
Nutrient agar plates
Ampicillin plates
Sterile swabs
Permanent marker
Antibiotic disks
Toothpicks
Live cultures, *E. coli*, *B. subtilis*
Crystal violet stain
Gram's iodine stain
Decolorizer stain
Safranin stain
Slides
Coverslips

---

# INTRODUCTION

---

## Studying Microbial Cultures

Because microbes are so small, studying them in the lab usually requires growing large numbers of them. Bacteria are typically grown in nutrient-rich liquid broth, or in Petri dishes ("plates") containing the same broth made solid by the addition of a gelling agent (usually agar, which comes from seaweed). It is estimated that fewer than 1% of all bacterial species can be grown in the lab. Those that cannot be grown either have unusual nutritional requirements that are not known or require the presence of other organisms with which they live in close association.

For bacteria that do thrive in the lab, it is easy to obtain many cells—a single *Escherichia coli* cell added to a milliliter of liquid broth will produce more than 100 million cells overnight! When grown overnight on solid medium in a Petri dish, a single *E. coli* cell will produce an entire colony of cells several millimeters across. Microbial samples from the environment can be cultured by depositing cells onto nutrient agar plates. The resulting colonies can be studied further using various biochemical tests. These tests are often useful in classifying bacteria, and are very frequently used in medical settings to determine the type of bacteria causing an infection.

The most common biochemical test used to classify bacteria is **Gram staining**, named after its inventor, Hans Christian Gram. Certain bacteria are turned purple by the Gram-staining procedure and therefore are called **Gram positive**. Bacteria that do not turn purple (and instead remain a pinkish color) are called "**Gram negative**. The distinction is important because Gram-positive and Gram-negative bacteria tend to be sensitive to different types of antibiotics. In some critical medical situations, Gram staining of body fluids is used for a quick diagnosis. For example, spinal fluid might be examined by Gram staining if bacterial meningitis is suspected. Classification of the infecting bacteria can help doctors to choose an appropriate treatment.

The structure of the cell wall determines whether a bacterium is Gram positive or Gram negative (see Figure 1). Bacterial cells are surrounded by **peptidoglycan**, a large molecule made up of sugars and amino acids. Gram-positive bacteria have a thick layer of peptidoglycan, whereas Gram-negative bacteria have a thin layer of peptidoglycan that is surrounded by an outer membrane composed of lipids and sugars.

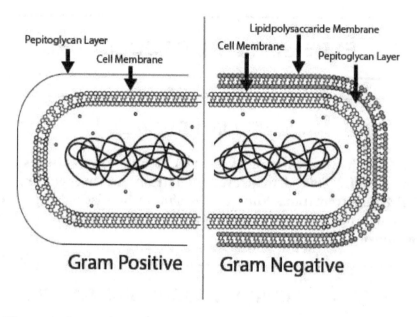

**Figure 1** Comparison of Gram-positive and Gram-negative bacteria.

The most well-studied Gram-negative bacterium is *E. coli*. The most well-studied Gram-positive bacterium is *Bacillus subtilis*, which is harmless but closely related to the anthrax-causing *Bacillus anthracis*. When viewed using a microscope, *E. coli* and *B. subtilis* look almost identical (see Figure 2). Thus, we need to use a biochemical test such as Gram staining to tell them apart.

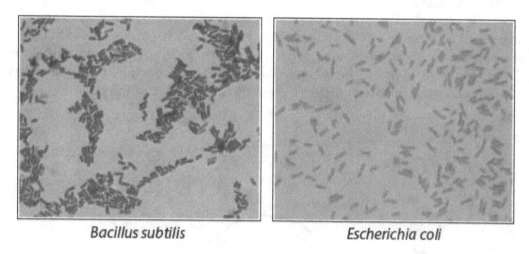

**Figure 2** Microscope images of *B. subtilis and E. coli*.

The presence or absence of the peptidoglycan layer can also affect the sensitivity of bacteria to antibiotics. Some antibiotics kill Gram-positive bacteria but not Gram-negative bacteria, while other antibiotics do the reverse. In this lab, you will investigate whether certain antibiotics are **"selective** for Gram-positive or Gram-negative bacteria or whether they are **broad spectrum** drugs that kill both types of bacteria.

# EXPERIMENTAL PROCEDURES

---

### SAFETY INFORMATION

<u>CAUTION</u>: **You must wear gloves when handling live cultures of *E. coli* and *B. subtilis*. This is to protect you from the live cultures and to minimize contamination. Your gloves should be discarded in the large biohazard box at the front of the room. Also gloves should be worn when wiping down bench with ethanol.**

---

## PART A: PREPARING MICROBIAL SAMPLES

1. You want to minimize the introduction of additional microbes to the agar plates. Before starting the experiment, it is important that you clean your lab bench using an ethanol solution.

2. Retrieve 8 nutrient agar plates from the empty lab bench. You will use 4 of these agar plates to culture samples from collection sites and 4 of these plates to experiment with live cultures and antibiotic disks.

3. Using a marker, label around the edges of the bottoms of a set of 4 plates with the collection sites listed below. **You must also label the plates with your initials and lab section number. Refer to Figure 3 for proper labeling. It is important that you clearly label your plates so you are able to identify them during Week 2 of the lab.**

   a. Bottom of shoe
   b. Mouth
   c. Front of cell phone
   d. Bathroom

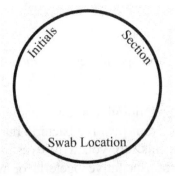

**Figure 3** Labeling of Part A plates.

4. Open a sterile swab package. **Take care not to touch the cotton end with your fingers. Use a fresh swab for each culture. Immediately dispose of the swabs in your bench top biohazard bag after streaking each plate.**

5. Wipe the cotton tip of the swab all along the surface being tested. Rotate the swab in a circular motion during collection to cover all sides with potential microbes.

6. Once you have picked up sufficient amounts of bacteria on the swab, you will transfer the organisms onto the corresponding labeled nutrient agar plate. The goal is to create an even layer of bacterial growth on the surface of the agar. This is known as a **bacterial lawn. To prevent cross-contamination for the Bathroom plate, bring the plate into the bathroom and swab there.**

7. Evenly streak the swab over the entire agar surface of the plate. The agar is a consistency similar to Jell-O. Do not press too hard or you will puncture the agar gel. Rotate your swab, and spread the bacterial growth in various directions to ensure the area is completely covered. Use Figure 4 below as a guide.

**Figure 4** Illustration of an evenly streaked plate.

8. Repeat the streaking procedure for all of your collection sites, and place the lids on the plates.

9. Select a sample that you collected from one of your environmental sites. Swab the sample on the surface of an ampicillin plate.

10. Once finished, place plates from Part A aside before starting Part B.

# PART B: STUDYING THE EFFECT OF ANTIBIOTICS

## Testing Three Antibiotics

1. Now you will prepare 2 plates using prepared cultures of *E. coli* and *B. subtilis*. In addition to streaking cultures on the plates, you will also be testing the effects of various antibiotic disks. Divide each plate into three zones as pictured in Figure 5. Label around the edges of the bottom of the plates using the marker as shown.

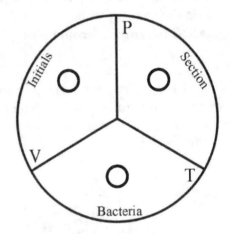

**Figure 5** Labeling of Part B plates.

2. The tubes of prepared cultures contain a slanted surface of agar with the live culture growing on top of the slant. Using a cotton swab, swab the top surface of the agar (do not pierce the agar) to collect the live culture.

3. Using the same technique you did for the last section, streak the prepared cultures of *E. coli* and *B. subtilis* onto the labeled agar plates to create bacterial lawns.

4. You will now add three disks containing antibiotics to the three zones of your agar plates. The three antibiotics that you will be testing are penicillin, vancomycin and tetracycline. **Label the zones P, V, and T.**

5. The antibiotic disks are stored in cartridges that require the use of a dispenser, which is already attached to each cartridge. Hover the bottom of the cartridge/dispenser apparatus over the middle of the correctly labeled zone. Squeeze the bottom of the dispenser to release the disk. Your instructor can demonstrate the proper technique if you are having difficulty.

6. Your plates will be stored upside down in the incubator to prevent condensation from dripping on the agar. Because of this, it is important that your antibiotic disks are secure in the agar. You may need to use a toothpick and apply light pressure to the disk. Please note that the toothpicks are not sterile and should only touch the disk, not the agar plate. Use a fresh toothpick each time.

7. Once all three disks have been dispensed onto the plates, place the lids on the plates, and set them aside.

## Investigating New Antibiotics

1. Using the same *E. coli* and *B. subtilis* cultures, you will now prepare two additional plates. Divide your plates into three zones.

2. You will now choose three additional antibiotics from the choices provided. Label the three zones of your plates according to the first letter (or first two letters) or the antibiotic's name (**make sure each label is unique**). Dispense the three disks of your choice in the appropriate zones following the same set of instructions as the previous section.

3. Record which antibiotic disks you selected for the experiment on the Data Sheets.

### PREPARING PLATES FOR INCUBATION

If you have not already done so, dispose of all used swabs in the desktop biohazard bag, and **initial the bottoms of all your plates with your name and lab section to identify which plates belong to your group.** Your instructor will transfer the plates into an incubator set at 37°C. These plates will incubate over a week's time and will be available for further testing during your next lab session.

# PART C: THE GRAM STAIN

In this part of the lab, you will be generating a Gram stain of selected organisms from your prepared cultures. This test is a major tool in the field of biology and is a primary method for both clinical and academic identification of microorganisms. It is used to detect the biochemical makeup of the cell wall in bacteria and other microorganisms. You will stain one of the organisms (*E. coli*, *B. subtilis*), and your partner will stain the other. The Gram staining procedure is summarized in Figure 6.

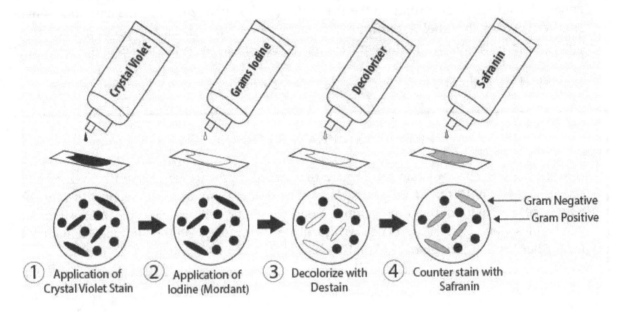

**Figure 6** Gram-staining procedure.

## Sample Preparation/Heat Fixing

1. Take a fresh blank slide, put it on the staining rack, and add **4 drops** of distilled water to the surface of the slide.

2. Take a sterile swab, and gently roll the swab in a circular motion within the tube containing the bacterial sample. Make sure that you cover all sides of the swab with bacteria (as done in Part B).

3. Once you have obtained enough of the bacterial sample, gently mix the tip of the swab in the distilled water. This will create a smear, and the water should become cloudy after the addition of your sample.

4. Turn on the hot plate to **200°C**, and wait **5 minutes** for it to heat up.

5. Once the plate is hot, hold your slide on one side with forceps, and place it on the plate, making sure the smear is fully on the plate.

6. Watch the slide and wait until all the water has been evaporated. This process will take approximately 20 to 30 seconds. **Do not leave the slide on the plate after evaporation or else you will burn the sample.**

7. After the water has evaporated, remove the slide from the plate with forceps, and place it on the staining rack to cool. **Be careful—the slide will be very hot!**

## Staining

1. Cover the slide with the **crystal violet stain**. Let it stand for 60 seconds.

2. Pour off the crystal violet stain by tipping the slide, and gently rinse the excess stain with a stream of distilled water from the plastic water bottle into the staining tray. **The objective of this step is to wash off the stain, not the fixed culture.** Shake off any of the excess water into the tray.

3. Cover the slide with **Gram's iodine**. Let it stand for 60 seconds.

4. Pour off the iodine solution, and rinse the slide with distilled water. **The objective of this step is to wash off the stain, not the fixed culture.** Shake off the excess water into the tray.

5. Add **10 drops** of **decolorizer,** and let the solution trickle down the slide for 5 to 10 seconds. The exact time to stop is when the solvent is no longer colored as it flows over the slide. Further delay will cause excess decolorization in the Gram-positive cells, and the purpose of staining will be defeated.

6. Counterstain the smear by covering the slide with **safranin**. Let it sit for a final 60 seconds.

7. Wash off the safranin with distilled water. Blot with slide paper gently to remove the excess water. Make sure not to obstruct the smear while drying.

8. Once dry, add a cover slip on top of the final smear, and observe it under the microscope at the station located on the back bench.

9. The cell walls of **Gram-positive** bacteria will retain the first dye and appear violet. By contrast, **Gram-negative** bacteria will lose the color of the first dye and appear red from the counterstain. Record your results in **Data Table 1** on the Data Sheets. Share your results with your partner, and record the results from the other bacterial sample in **Data Table 1**.

# CLEAN-UP CHECKLIST

It is essential that you clean up properly after the experiments are complete.

❑ Use ethanol to thoroughly clean your lab bench.

❑ Dispose of all swabs in the benchtop biohazard bags. Empty the swabs from this lab into the large biohazard box once you're finished with swabbing, and return the bag to your bench.

❑ Dispose of used gloves in the large biohazard box at the front of the room.

❑ Place plates upside down on an empty lab bench.

---

# DATA SHEETS:
# Bacteria and Antibiotics (Part 1)

---

Name: _____ Date: _____

Lab Instructor: _____ Section: _____

## PART C: PREPARING MICROBIAL SAMPLES

Antibiotic 1: _____

Antibiotic 2: _____

Antibiotic 3: _____

## PART C: THE GRAM STAIN

### Data Table 1: Gram-Staining Results

| Name of Organism | Color of Cells | Gram Reaction (+/−) |
|---|---|---|
|  |  |  |
|  |  |  |

Sketch your observations in the space provided below. Use **colored pencils** to represent the colors of the bacterial colonies.

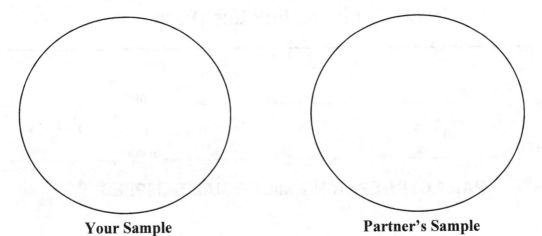

**Your Sample**                    **Partner's Sample**

# EXPLORATION 21B
# Bacteria and Antibiotics (Part 2)

## Overview

During the second part of this lab project, you will carefully examine the results of your experiments from last week.

## Preparation

You should be familiar with the following concepts and techniques:

- How penicillin kills bacteria
- Two mechanisms by which antibiotics kill bacteria
- Experiment used to determine whether an antibiotic is effective

## Equipment and Supplies

**Equipment**
Ruler
Plexiglass

**Supplies**
Colored pencils
Permanent marker
Label tape
Ethanol, 50%

# INTRODUCTION

## How Do Antibiotics Work?

How do antibiotics prevent the growth of bacteria? Are some antibiotics more effective than others against certain types of bacteria? What factors determine whether a specific bacterium is susceptible to a particular antibiotic? These are just three of many questions we can ask about how antibiotics work.

The term **antibiotic** is used to describe a class of drugs that inhibit the growth of bacteria However, not all antibiotics work in the same way. The most famous antibiotic—**penicillin**—kills bacteria by interfering with the synthesis of the bacterial cell wall. The loss of structural integrity in the cell wall causes the bacterial cell to rupture and spill the fluid cytoplasm from inside the cell. Other antibiotics interfere with the ability of the bacterial cell to synthesize proteins in molecular "protein-making factories," called **ribosomes**. When this happens, the bacteria do not have the essential proteins they need to survive or reproduce. Yet another type of antibiotic targets the DNA of the bacterial cell, which prevents bacterial cells from replicating properly when they divide and grow. These three general methods of antibiotic action are shown in Figure 1. There are also other ways in which antibiotics attack bacterial cells, but these three are the most common.

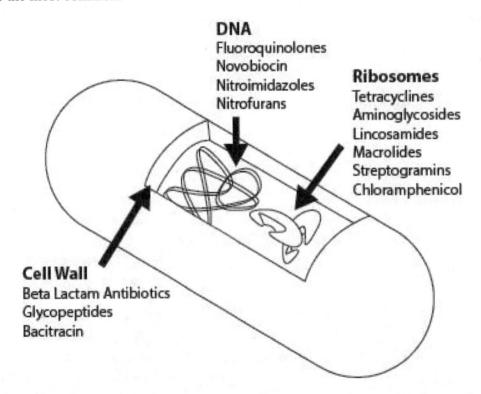

**Figure 1** Major targets of common antibiotics drugs. A list of drugs is shown below the region of the cell that the drug targets.

How can we tell if an antibiotic prevents the growth of bacteria? In last week's experiment, you placed antibiotic disks on an agar plate that was streaked with bacteria. If the antibiotic halts the growth of bacteria, there will be an **inhibition zone** around the antibiotic disc. Figure 2 shows the effect of different antibiotics on a bacterial sample. Note the variation in the size of the inhibition zone around each antibiotic disk. For some antibiotics, the inhibition zone is large, indicating that the antibiotic is very effective at preventing bacterial growth. For other antibiotics, the inhibition zone is very small, indicating that the antibiotic had little effect on the bacteria. By comparing the sizes of the inhibition zones in your experiments, you will be able to evaluate the effectiveness of different antibiotic drugs in prevent the growth of *Escherichia coli* and *Bacillus subtilis*.

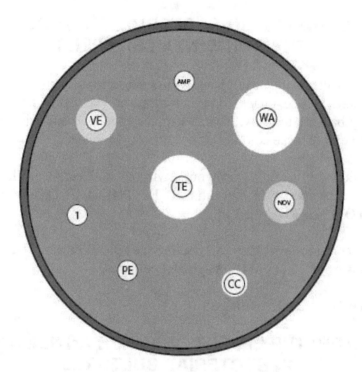

**Figure 2** The effectiveness of an antibiotic drug against a particular type of bacteria is measured by the size of the inhibition zone around the antibiotic disk. A larger inhibition zone means that the antibiotic is more potent against the bacteria.

---

# EXPERIMENTAL PROCEDURES

---

<u>**SAFETY INFORMATION**</u>

<u>CAUTION</u>: **Gloves must be worn for the duration of the experiment.**

---

## PART A: ANALYZING BACTERIAL CULTURES FROM ENVIRONMENTAL SAMPLES

Last week, you streaked four agar plates with swab samples that you collected from four environmental locations. These plates were placed in an incubator set at 37°C, the optimal temperature for many bacteria to grow. If your samples contained bacteria that can grow on agar plates, you will be able to observe colonies of bacteria on the plate.

1. Study the plates that you streaked from your environmental samples. **The agar plates do not need to be opened to observe the results. KEEP THEM CLOSED and view the results from the BOTTOM OF THE PLATE.**

2. Sketch your observations in **Question 1a** on the Data Sheets. **Use colored pencils to represent the observed colors of the bacterial colonies**.

3. Answer **Question 1b** on the Data Sheets.

## PART B: THE EFFECT OF THREE ANTIBIOTICS ON BACTERIAL CULTURES

1. Examine the set of plates prepared last week containing the antibiotics **penicillin (P), vancomycin (V), and tetracycline (T)**. Sketch your observations in **Question 2** on the Data Sheets, and show the inhibition zone around each disk.

2. Take a closer look at the inhibition zone surrounding each antibiotic disk. **Use the ruler to measure the diameter of each inhibition zone in centimeters, NOT inches,** as seen in Figure 3. Record your data in **Question 3** on the Data Sheets.

3. Answer **Question 4** on the Data Sheets. Use the guide presented with that question to classify each zone as +++ (large), ++ (medium), + (small), or − (does not exist).

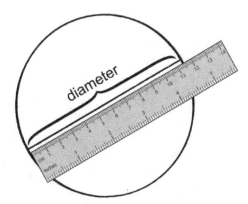

**Figure 3** How to measure the diameter of the zone of inhibition.

## PART C: THE EFFECT OF ADDITIONAL ANTIBIOTICS ON BACTERIAL CULTURES

1. Study the set of plates containing the three antibiotics of your choice. Sketch your observations in **Question 5** on the Data Sheets, and show the inhibition zone around each disk.

2. Take a closer look at the inhibition zone surrounding each antibiotic disk. Use the ruler to measure the diameter (in cm) of each inhibition zone. Record your data in **Question 6** on the Data Sheets.

3. Answer **Question 7** on the Data Sheets. Follow the guidelines presented with that question to classify each inhibition zone.

4. Once you have determined the inhibition zones for your antibiotics, record your data to share with the rest of the class.

5. Answer **Questions 8 and 9** on the Data Sheets.

---

# CLEAN-UP CHECKLIST

---

It is essential that you clean up properly after the experiments are complete.

❏ Tape the used agar plates shut.

❏ Dispose of your taped agar plates in the biohazard box at the back of the room.

❏ Use 50% ethanol to clean your lab bench.

```
┌─────────────────────────────────────────────────────────┐
│                   DATA SHEETS:                            │
│          Bacteria and Antibiotics (Part 2)                │
└─────────────────────────────────────────────────────────┘
```

Name: _____ Date: _____

Lab Instructor: _____ Section: _____

## PART A: ANALYZING BACTERIAL CULTURES FROM ENVIRONMENTAL SAMPLES

**Question 1:** (a) Sketch your observations of the culture plates in the spaces provided below. Use **colored pencils** to represent the colors of the bacterial colonies.

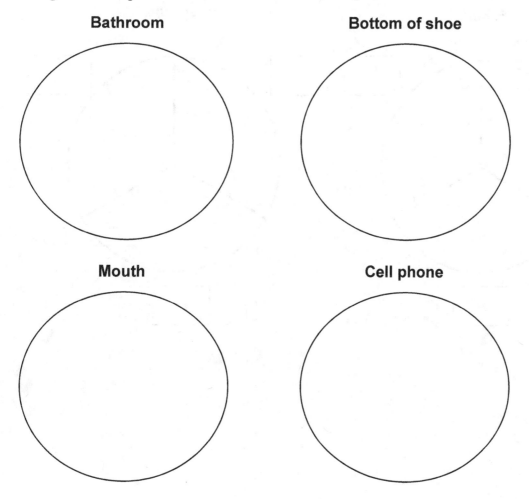

**Bathroom**

**Bottom of shoe**

**Mouth**

**Cell phone**

(b) Which environmental sample generated the largest number of bacterial colonies?

## PART B: THE EFFECT OF THREE ANTIBIOTICS
## ON BACTERIAL CULTURES

**Question 2:** In the boxes provided, label which antibiotic disk (P, V, T) was disposed into each zone. Sketch your observations of the culture plates in the spaces below. Your diagrams should show the antibiotic disks and the **relative size of the inhibition zone around each disk.**

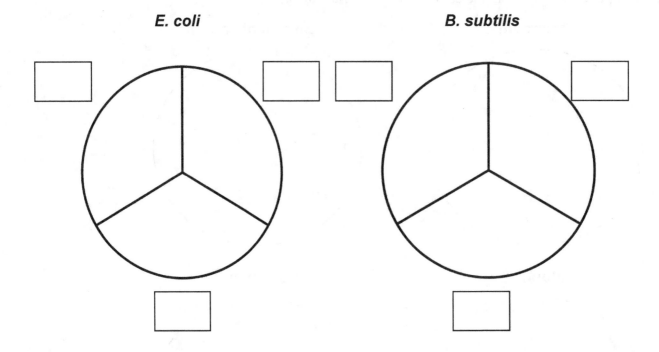

**Question 3:** Measure the diameter (in cm) of each inhibition zone, and record your results in Data Table 1 below.

**Data Table 1: Inhibition Zones of Penicillin, Vancomycin, and Tetracycline**

| Antibiotic Type | *E. coli* | *B. subtilis* |
|---|---|---|
| Penicillin | | |
| Vancomycin | | |
| Tetracycline | | |

**Question 4:** Based on your measurements, complete Data Table 2 using the following key to indicate the **relative amount of inhibition** for each antibiotic.

| | | |
|---|---|---|
| >3 cm | + + + | Large zone of inhibition |
| 2–3 cm | + + | Moderate zone of inhibition |
| 1–2 cm | + | Small zone of inhibition |
| <1 cm | -- | No zone of inhibition observed |

**Data Table 2: Relative Amount of Inhibition of Penicillin, Vancomycin, and Tetracycline**

| Antibiotic Type | *E. coli* | *B. subtilis* |
|---|---|---|
| Penicillin | | |
| Vancomycin | | |
| Tetracycline | | |

## PART C: THE EFFECT OF ADDITIONAL ANTIBIOTICS
## ON BACTERIAL CULTURES

**Question 5:** In this experiment, **you selected 3 antibiotics** to investigate. In the boxes given below, write the name of the antibiotic disk in each zone (don't be concerned if the name goes outside the box). Sketch your observations of the culture plates. Your diagrams should show the antibiotic disks and the **relative size of the inhibition zone around each disk.**

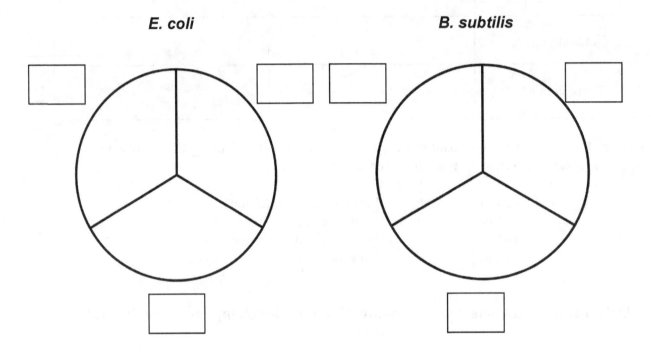

**Question 6:** Measure the diameter (in cm) of each inhibition zone and record your results in Data Table 3.

### Data Table 3: Inhibition Zones of Additional Antibiotics

| Antibiotic Type | E. coli | B. subtilis |
|---|---|---|
|  |  |  |
|  |  |  |
|  |  |  |

**Question 7:** Based on your measurements, complete Data Table 4 using the following key to indicate the **relative amount of inhibition** for each antibiotic.

| | | |
|---|---|---|
| **>3 cm** | + + + | Large zone of inhibition |
| **2–3 cm** | + + | Moderate zone of inhibition |
| **1–2 cm** | + | Small zone of inhibition |
| **<1 cm** | -- | No zone of inhibition observed |

### Data Table 4: Relative Amount of Inhibition of Additional Antibiotics

| Antibiotic Type | *E. coli* | *B. subtilis* |
|---|---|---|
| | | |
| | | |
| | | |

**Question 8:** When you have completed Data Table 4, add the summary of your observations to the "master list" that compiles all the data from the entire lab section. Answer the following questions based on the complete set of data:

(a) What **three antibiotics from Part C** were **most effective** against *E. coli* bacteria?

1.

2.

3.

(b) What **three antibiotics from Part C** were **most effective** at against *B. subtilis* bacteria?

1.

2.

3.

(c) Are there any antibiotics that appear on **both lists**—that is, drugs that are effective at killing **both** *E. coli* and *B. subtilis*?

**Question 9:** Based on your observations in this experiment:

(a) Which antibiotic would you prescribe to treat an infection by **Gram-negative bacteria**? Explain your choice.

(b) Which antibiotic would you prescribe to treat an infection by **Gram-positive bacteria**? Explain your choice.